"I have tried to weave my own am[...]
love and grief, longing and anger, [...]
ing it clear that there is no perfecti[...]
to defend but much to use and much to va[...].

—Mary Catherine Bateson,
With a Daughter's Eye

"A loving and sympathetic personal memoir. . . . Stunning revelations." —*Newsweek*

"Hers is a single voice, speaking honestly, eloquently, and sometimes wryly about her famous parents. . . . An invaluable addition to the record, and above all, a profoundly human document in which . . . personal experience illumines larger issues. . . . Splendid."
—*Science*

"There will be many other studies of the life of Margaret Mead and Gregory Bateson; all will be the more comprehending because of their daughter's portrayal of them—analytical, vivid, and affecting."
—*Los Angeles Times Book Review*

"Bateson has an exquisite, mature, and, when necessary, unsparing self-consciousness. She is as honest about her own moments of pain and disappointment as she is grateful and admiring."
—Judith Thurman, *Vogue*

"*With a Daughter's Eye* gives us the best account of Bateson's father, Gregory Bateson . . . an original thinker whose ideas are only now beginning to be appreciated. . . . A charming book."
—Ashley Montague, *New York*

"[Dr. Bateson] understands the originality of her parents' formidable minds, conveys the qualities of their remarkable personalities, and shows them as entertaining in intimacy. . . . Her memoir is a triumph."
—*The New Yorker*

With a
Daughter's Eye

With a Daughter's Eye

A Memoir of Margaret Mead and Gregory Bateson

MARY CATHERINE BATESON

Perennial

An Imprint of HarperCollins*Publishers*

Grateful acknowledgment is made for permission to quote the following:

"When Earth's Last Picture Is Painted" and "The Son of Martha" from *Rudyard Kipling's Verse;* Definitive Edition reprinted by permission of the National Trust and Doubleday & Company, Inc.

"Recuerdo" and "First Fig" by Edna St. Vincent Millay by permission of Mrs. Norma Millay Ellis.

"Spring is like a perhaps hand" by e.e. cummings copyright 1925, © 1973, 1976 by Trustees for the e.e. cummings Trust.

Blackberry Winter: My Earlier Years by Margaret Mead by permission of the Institute for Intercultural Studies.

Cecil Beaton photograph courtesy *Vogue,* copyright © 1969 by Condé Nast Publications, Inc.

First HarperPerennial edition published 1994.
First Perennial edition published 2001.

Designed by The Book Design Group / Matt Perry Ratto

Library of Congress Cataloging-in-Publication Data is available.
ISBN 0-06-097573-3

01 02 03 04 05   WB/RRD   10 9 8 7 6 5 4 3 2 1

For my Aunt Marie

Contents

Acknowledgments

Before I thank those who contributed directly to this work, it seems important to thank those who would have been happy to help had I called on them, family members, friends, and colleagues of my parents, and the archivists who have worked with their papers at the Library of Congress and the University of California at Santa Cruz. I am well aware that had I pursued the paths of interviewing and archival research this would have been a richer and no doubt a more accurate book, but I have chosen to concentrate on my own perception and experience. The errors are also my own. I have made my choices with a sense of gratitude, however, for help that would gladly have been given. I am grateful too for occasional illuminating conversations that might have been carried further, had I pursued them, especially with Lois Bateson, Geoffrey Gorer, Madeline Lee, and Rhoda Metraux, and conversations about the project with my father before his death.

The person to whom I turned most often with questions

during the writing of the book was Marie Eichelberger, who died at the end of 1983, knowing that the book would be dedicated to her but without having seen the text. At a crucial moment in the writing, when I felt I needed some input from readers of my parents' generation, close both to them and to me, and close to both personal and intellectual issues, I turned to Erik and Joan Erikson, asking them to read as "advocates and surrogates." I am grateful to them for their specific comments, but above all for emphasizing their belief that what my parents would have wanted me to do was to speak honestly in my own voice, out of my own feelings.

Places and institutions have been important in the completion of this book, which was begun in 1979. My most intensely productive writing time was a month spent in Italy in the summer of 1983 at the Bellagio Conference and Study Center of the Rockefeller Foundation. The 1983–84 academic year was spent as a Fellow at the Mary Ingraham Bunting Institute, the research institute for women of Radcliffe College, and it is there that I have completed the book. Both places were important as much for the conversations that took place with other scholars as for hospitality and facilities. The 1983–84 year of work was made possible by a leave granted by Amherst College. During this year, a number of colleagues and friends have contributed or helped in important ways. In this context, I would like to thank Ellen Bassuk, Robert Bezucha, Deborah Gewertz, Linda Gordon, Barbara Kreiger, Alan Lelchuk, Linda Perry, Roy Rappaport, and David Sofield. I also owe thanks to William Irwin Thompson, to my editor, Pat Golbitz, to Barbara and Frederick Roll, and to John Brockman. The photographs in this book have been selected to rely as much as possible on the work of photographers who were friends and professional associates of my parents. I want especially to acknowledge (in chronological order of photographs used) Jane Belo, Karsten Stapelfeldt, Paul Byers, Ken Heyman, and Frederick and Barbara Roll.

One should perhaps acknowledge all sources of important influence, even if adverse. Thus, without the Iranian revolution, this book would probably not have been written; without Amherst College it would have been written sooner, more peacefully, and perhaps more superficially; without Derek Freeman I would have lacked a spice of anger that has strengthened some of my affirmations.

The person who has been most closely and constantly concerned in this writing, at every level, is my husband, J. Barkev Kassarjian; his advice and support have made this one of the creative pleasures of our marriage. Our daughter, Vanni, has helped too—most of all by reinforcing the conviction I bring from my childhood that being a parent and being a daughter, while not always happy or easy, are deeply joyful.

The Mary Ingraham Bunting Institute
Cambridge, Massachusetts
June 1984

Preface to the 1994 HarperPerennial Edition

When I committed myself in 1979 to write a memoir of my parents, I did so with scant reflection on the psychology of the experience that lay ahead or the nature of the genre in which I planned to work. The notion of writing about my parents as I knew them seemed intuitively obvious to me, just as the commercial viability of a daughter's memoirs of two famous parents seemed obvious to the publishers. As a result, I grappled rather naively with issues that have been the subject of considerable critical exploration, issues that have continued to be important in my subsequent work with first-person narratives of other kinds. The writing allowed me to discover for myself an epistemological stance that has been reaffirmed repeatedly in feminist writings, the anchoring of observation and report in personal experience.

The process of remembering had some unanticipated effects. It was both more painful and more rewarding than I had expected. On the one hand, it brought a kind of closure. What I wrote became for me canonical—the memories I chose to include are the ones that stayed with me, the photographs in this book are now the images in my inner eye. Yet in other ways, the writing initiated forms of reflection that still continue, allowing me to become aware of the layered quality of a child's perception of parents, so that new stages in my life continue to open up aspects of my parents I was not ready to notice a decade ago.

Although we often distinguish between word and deed, linguists and philosophers have pointed out that words spoken and written are themselves deeds. Some utterances even create the reality they report. This is most easily recognized with utterances that include a group of words called performatives. When I begin a sentence with "I swear," or "I admit," it has been said, the swearing or the admission commits me in new ways. Speech acts such as "I now pronounce you man and wife" go even further. Yet one can imagine an unstated performative framing whatever we say, for each expression or report makes our world a little different.

A memoir is framed in just such a way. Even as I naively saw myself telling a story, I was making choices: asserting, claiming, acknowledging. This narrative about my parents was what I chose to say about where I come from, full of implications about who I am and what I expect to become. Putting a version of the past on record is an act of decision and affirmation, setting commitments for the future. Some of that future is now past, but that is a different story, and besides, the wench is dead: I am not the same person I was then. This is why I have decided to make no changes or additions in a memoir that took shape at a particular moment in my own life. There has always been much that could have been included, but this book never pretended to completeness. If I were to write it today I would write it differently, but that would be a completely different book.

As far as the public understanding of my parents' work is concerned, I continue to hope that this book will provide a point of entry, but I go on struggling to avoid becoming an arbiter of orthodoxy in interpretations of their ideas. I have been chary of monuments and focused my efforts on making their work available to those who want to study it and reach their own conclusions. Their living is another matter. Because my parents, especially my mother, were pioneers in their understandings of family life, an account of what they did and how it affected my growing up may be valuable as we work toward new understandings of the family. The narrative stands virtually alone as an account of a child's experience of sustained affection and relationship with both parents after a divorce. In all the correspondence I have received since the book appeared, the theme that I have found most moving is the use of Margaret and Gregory's experience as inspiration for evolving new patterns of relationship within and between families, giving fresh and flexible meaning to the term "family values." Part of the force of this memoir is the assertion that Margaret and Gregory were, together, effective parents, complementing each other, reunited in these pages.

Inspiration, but not models, for I believe deeply that we must all compose our lives without relying on single role models. Margaret and Gregory lived lives that were closely related to the times in which they lived, lives which cannot be emulated today. I am often asked, what did you choose to do differently? Many things. But of course my parents do not represent only their own choices; they represent the possibility of choosing, which I still affirm.

With a
Daughter's Eye

I

Prologue: The Aquarium
and the Globe

My parents, Gregory Bateson and Margaret Mead, were scientists and teachers, not only in the wider community in which they worked and published, each becoming famous in different ways and touching many lives, but in the domestic circles of family and friendship as well. For them, the intimate was projected onto the widest screen, even as knowledge from far places was worked into the decisions of everyday life. The minds of both sought patterns of completeness, wholes, and so they thought of worlds entire, whether these worlds were minute images of microscopic life within a drop of water or the planet wreathed in cloud.

They thought of worlds and drew me into them. There were worlds to be built and worlds to be imagined, worlds to be held and cherished in two hands and worlds of abstract argument, in spherical tautology. The small primitive societies in which each did ethnographic work were worlds of one kind, complete communities to be described and understood, but along with these there was the challenge to construct and be responsible for the wholeness of family, a world for a child to grow in, a biosphere to protect, the possibility of the bright sphere shattered. Growing up was a passage from the microcosm, a motion through concentric metaphors. Even in the smallest of shared spaces a camera or a notebook stood for a possible opening up to the macrocosm.

A child moves out through concentric worlds even with her first steps, but whether these worlds are encountered as wholes or as fragments and whether they provide an entry to other spheres of imagination and experience depend on how they are presented, how attention is gradually shaped and the cosmos gradually unfolded.

In Holderness, New Hampshire, where we spent many summers, a long field runs down toward the lake. At the bottom, just short of the strip of woods that shields the shore, there lies a broad patch of springy moss, like a bright green eiderdown spread out under the trees. This was a place my mother had picked to be alone with me in counterpoint to the large household in which we stayed. We used to wander there for an hour or so, especially in the early morning. Sometimes we found spiderwebs stretched flat above the moss between protruding grass stems, with dewdrops still shining on them. These she showed to me as fairy tablecloths, the damask spun by tiny fingers, with crystal goblets and silver plates still spread out, for the feckless fairies went off to sleep at dawn without cleaning up. Then she showed me red-tipped lichens as small as a pinhead—fairy roses—and searching along the ground we found their serving bowls, the bases of acorns.

My great-grandmother had taught my mother how to identify and draw all the plants of her Pennsylvania childhood, but for me the flowers had only colloquial names and were lenses of fantasy: Indian paintbrush, black-eyed Susan, milkweek, Jack-in-the-pulpit. "I know," she sang, "where the fringed gentians grow."

My father had the English habit of latinizing in the woods or in the garden. The intricacies he showed me between the grass stems were of another sort, perhaps a beetle or a moth living out quite different dramas. When I look at the field with his eyes, I see it as a series of complex symmetries and relationships, in which the position of the spiderweb above the moss hints at the pathways of foraging insects. The petals of daisies can be used to count—"He loves me, he loves me not"—because they are not true petals but flowerets—otherwise their number would be set in the precise morphology of the flowering plants.

"Once upon a time," my mother would narrate as the sun moved higher in the sky, "in the kingdom between the grass stems, there lived a king and a queen who had three daughters. The eldest was tall and golden-haired and laughing, the second was bold and raven-haired. But the youngest was gray-eyed and gentle, walking apart and dreaming." The story varies but the pattern remains the same, woven from the grass of the meadow and the fears and longing of generations. For this king and queen lived in no anarchic world, but in a world of rhythm and just symmetries. Their labors, quests, and loves grew out of each other with the same elegance that connects the parts of a flowering plant and its cycles of growth. At their court, as at the fairies' banquets, crystal goblets and courtly etiquette reflected a social order. Prince and princess find one another in a world of due peril and challenge and happiness ever after. The flower is pollinated, seed is formed, scattered, and germinated. Look! The silk in the milkweed pods is what the fairies use to stuff their mattresses. Blow on the dandelion down to make a wish, anticipating the wind. Pause in the middle of fantasy to see the natural world as fragile and precious, threatened as well as caressed by human dreaming.

Worlds can be found by a child and an adult bending down and looking together under the grass stems or at the skittering crabs in a tidal pool. They can be spun from the stuff of fantasy and tradition. And they can be handled and changed, created in little from all sorts of materials. On a coffee table in the center of our living room, which often held toys and projects of mine, I constructed a series of worlds on trays. One of these was meant to depict a natural landscape, built up from rocks and soil, with colored sand and tinted strawflowers set into it. Another was inspired by a book my father had read to me in which a child constructs a city with cups, dishes, and utensils from the kitchen and then visits it in his dreams. My mother, in that same period, was fascinated by the World Test of Margaret Lowenfeld,[1] an English child analyst. This projective test consists of a tray of moist sand and a vast array of miniatures: people and animals, trees and houses and vehicles. In using the test, one molds the soil and handles the objects, arranging and changing them, and then weaves narratives within the world one has created, so that the creation of a microcosm becomes the expression of an inner, psychic world, a world that embodies pain and perplexity as well as symmetry.

The other kind of world that I constructed as a child was represented by a series of aquaria set up with my father. An aquarium is bounded, like a city or a landscape built on a tray, but the discipline that goes into building it is different, for it is alive. In the fantasy world, the discipline is primarily aesthetic: Here is the forest and here the open valley; here the dragon lurks and here the river runs. In an aquarium it is necessary to balance the needs of living creatures and their relationships with each other, the cycles of growth and respiration and decay. Here among the thicker water plants, newly spawned swordtails shelter lest they be devoured. The snails that move sedately on the glass control the algae, and on the sandy bottom catfish prowl continually, scavenging the pollution of living that never occurs in fairy tales.

It is not easy to give a child a sense of the integrity of the biosphere. Even today there seem to be few who see themselves as living within and responsible to a single interconnected whole. As a very small child, asked what I wanted for Christmas, I am supposed to have answered that I wanted the world, and my parents gave me a globe. I do not know now whether I found the hollow painted sphere a very satisfactory present. I remember it standing at one end of the long living room for years, next to the aquarium, and yet I am sure that none of us in those days saw the two as metaphors, each of the other, a metaphor that we now can easily make through the mediating symbolism of the picture of the earth as seen from space.

Through my mother's writing echoes the question "What kind of world can we *build* for our children?" She thought in terms of building. She set out to create a community for me to grow up in, she threw herself wholeheartedly into the planning and governance of my elementary school, and she built and sustained a network of relationships around herself, at once the shelter in which I rested and the matrix of her work and thought. Not so my father, for the most complex actual worlds I knew him to set out to build have been aquaria and conferences, temporary constellations of people who learn to think in counterpoint to each other, moving

toward a unity of mental process. He was less free than my
mother to build and imagine, but I remember him for creating moments of attention when the patterned wonder of
some wild place or human interaction became visible.

The mornings of fantasy with my mother in a New
Hampshire meadow, the hours spent assembling an aquarium with my father, these are moments shared in the microcosm that will not be exactly duplicated in any other memory
but my own. But the writing of this book expresses a belief
that multiple small spheres of personal experience both echo
and enable events shared more widely, expressions of
moment in a world in which we now recognize that no
microcosm is completely separate, no tide pool, no forest, no
family, no nation. Indeed, the knowledge drawn from the life
of some single organism or community or from the intimate
experience of an individual may prove to be relevant to decisions that affect the health of a city or the peace of the world.

This is not a book about Margaret Mead or Gregory
Bateson that strives for completeness and objectivity and
attempts to define the place of one or the other in the wider
world; rather, it concerns the moments and modalities of
my relationship with each and occasionally, because these
informed the relationships, my sense of each with others or
of wider historical and professional contexts that converged
in me. Others have other stories to tell. My father, after his
divorce from my mother, had other children; my mother had
no other children but many godchildren; both had students
who knew them as teachers. As an adult I became an anthropologist myself and a colleague to each in their scientific
work, both like and unlike other colleagues. There will be
biographers aplenty who will attempt to synthesize multiple
views and wade through the published and unpublished doc-

uments that illuminate these lives, tracing the names and dates, evaluating and categorizing, but I have not wanted or dared to undertake these tasks. The wholeness of this book comes only from my experience and the effort to understand the first chapters of my own lifetime, incomplete, ambivalent, only partly conscious, involved as well in other worlds and relationships, a continuing odyssey through spheres of love and learning.

II
Baby Pictures

As I have mused back over memories of my relationships with my mother and with my father and of their relationship with each other as far as it was accessible to me, I have become more and more sharply aware of how many other lives are linked with these three, and of how the threads go back before my own birth in 1939 and continue after my parents' deaths in 1978 and 1980.

Margaret's story begins in a series of New Jersey and Pennsylvania farmhouses where her academic parents raised her and her younger siblings, a brother and two sisters. It continues through a year of college in the Midwest followed by a transfer to Barnard College. New York was Margaret's base for the rest of her life, on through graduate school at Columbia as a student of cultural anthropology under Franz Boas and an appointment at the American Museum of Natural History which she held for the rest of her life. By the time Margaret met my father she was already famous for her work in Samoa, and she had established a pattern of trips to the field and returns in which the work was written up not only as the careful documentation of exotic cultures in Samoa, New Guinea, and Bali but as relevant to the lives of ordinary Americans, to their decisions about how to raise their own children or order their public or private lives. She told that story herself in her autobiography, *Blackberry Winter*, hinting at a springtime with an edge of pain which makes possible the later harvest.

My father's story goes back through three generations of English academic life shaped by the traditions and expectations of Cambridge University and the scientific commitments of his father, a distinguished geneticist. Gregory was the youngest of three sons, but by the time he grew up both of his older brothers were dead. Only slightly younger than Margaret, Gregory was far younger professionally, fixed in the styles of academic bachelorhood, with only two slim articles in print and no clear sense of where he was heading as an anthropologist.

When Margaret and Gregory met, Margaret had already been divorced from her first husband, Luther Cressman, and was remarried to Reo Fortune, an anthropologist from New Zealand. Margaret and Reo met Gregory in New Guinea in 1932 while all three were doing research on the Sepik River, and after Margaret divorced Reo, she and Gregory married in 1935. The best years of their marriage were the years when they did fieldwork together in Bali and in New Guinea, years of such intensity that Margaret felt they had included a lifetime and produced as first progeny a family of books in which they shared in different ways.

All of this is history for me, known only at second hand, chewed over by now by a whole series of biographers. The first years of my own life are equally remote from my direct knowledge, and yet, clouded by forgetfulness, they remain most deeply and immediately present, the substratum of the feelings of later years. In trying to write about my relationship to my parents I feel a special need to recapture the period after my birth, for only in those years and only briefly did I experience them as parents together, before war and their estrangement intervened. I fill my imagination with the anecdotes and reminiscences that were told to me about my own childhood, and with photographic images, drawing on my sense of what lies deepest in my own personality and on the ways in which my own choices as a parent differ from the choices of my parents.

In some ways, the record is extraordinarily rich. Margaret and Gregory saw my childhood in a context they were committed to studying—child development and character formation as these vary in different cultural settings. All the societies in which they had done research had in common that they were relatively small and homogeneous, but they differed radically in the kinds of person who would be at home and productive in each. In their most important joint work, in Bali, they had worked closely together, Margaret taking notes and Gregory taking a vast number of photographs, with particular emphasis on the congruence between the experience of infants in their first few years and the wider structures of the society, the forms of cooperation and leadership, the shapes of the imagination in ritual and the arts.

They approached the experience of parenthood with the intention of questioning forms from their two backgrounds, combining elements of the British tradition from which Gregory came and the old American tradition from which Margaret came with care and discrimination, making innovations along the way. This meant that note taking and photography would go on and that my childhood would be documented and folded into their emerging understanding and later shared with many other people. In my family, we never simply live, we are always reflecting on our lives, and yet, against this background, as I write about my experience, I repossess it.

I start, as, I suppose, every child of divorce would, with the effort to see my two parents together in relation to me, to unite in one their separate images. The times in which I can see them both are few: a few years at the beginning of the war when I was an infant, one year after the war before they separated, transient professional meetings in the years that followed. Finally, as death approached, each one acknowledged illness and loss as occurring within a context they shared. In the last half year before my mother died, they met repeatedly, giving me again moments like the fragmentary images saved from my childhood: Gregory's skepticism modified by tenderness, Margaret's drive and energy once again expressed through a tiny body, the heaviness of her later years now wasted. The sweetness of these meetings reevoked for me hours of hovering, as a child, at the edge of intense conversations and explained to me why I did not retreat from abstraction and abstruse speculation. The play of intellect was a carrier of emotion, the conversation a form of love-making.

You could not see them together without thinking in terms of contrasts, and the same sharp sense of dissonance assails me when I look through old photographs or call up memories. Most striking was the difference in height and in their styles and rhythms of movement. My mother, barely five feet tall, was compact and economical in her motions, gathering all that she needed efficiently around her, reaching out from the elbow rather than from the shoulder. Gregory, just over six foot five, had spent much of his youth concealing his inches in a slouch and had more limbs and height than he knew what to do with. I visualize them now, sitting with me on a blanket on the ground outdoors. Gregory's legs are drawn up in peaks, one of his elbows resting on a knee, an *M* of angularity crowned with a lopsided *W*. Margaret sits on one hip with her feet drawn up and her skirt neatly arranged, like a lady riding sidesaddle, her hands gathered in her lap, but leaning forward in the intensity of speech. Her compactness created shelters: an enclosing lap that was stable to sit on, within encircling arms, or a very special space to cuddle next to her on the sofa, while she read aloud. Gregory's body, for a child, was like a jungle gym rather than a nest. The most glorious place to be was up on his shoulders, perched far above the crowd, ducking down to pass under doorways or the branches of trees.

Their rhythms were different also. Margaret was swift and sure of her intent as she moved through the day, almost as if she were following an agenda in which every activity was labeled, seemingly untiring but never wasting energy, ending phone conversations abruptly and rarely turning for additional farewells once a new trajectory was set. Gregory's day was filled with postponements and moments of sinking into quiescence, briefly aimless before all that length could be mustered for some next activity. His feet indeed were distant colonies, far from his caring; as he grew older they lost feeling, so winter and summer he wore his shoes without socks. Often his feet were left out of the warmth, protruding beyond the bedclothes and over the floor on beds too short for his height.

When I picture my parents I see their hands. Margaret's

were small and delicate with tiny half-moons in the nails, moving in symmetrical gesture with the palms cupped upward in front of her as she spoke, drawn back when the phrase was complete. Almost it seemed as if she were symbolically offering her breasts in her palms, a persuasion of nurture behind even the most trenchant argument. Gregory's hands were dramatic, long-fingered and angular, with large discolored nails. He used them asymmetrically in speech, and sometimes a hand would remain extended in gesture, forgotten. His maternal grandfather had been a famous surgeon at Guy's Hospital in London, whose students were said to be distracted from the details of surgical procedure by the beauty of his hands in motion. My grandmother, after passing on this family lore, remarked that Margaret had nice hands too, "though small," and I grew up watching my own hands for clues to who I might be becoming, moving hands imaged and reimaged in the poems I wrote as semaphores of intimacy or loneliness. My mother used to say that I used my hands like her in speech, but that in repose they resembled his.

I was born in New York in December 1939. Europe was already at war and my father, still a British subject, was groping for an appropriate role in the war effort. Always liable to fatalism and futility, he was surely less dubious then about the intentions and plans of others than he later became, but he always lacked a clear projection of the future. Margaret was expecting American involvement in Europe but the interval before Pearl Harbor gave her a fortunate period of leeway to "get life organized." She would have persisted indomitably in spite of uncertainty, constituting and reconstituting her arrangements to fulfill the multiple imaginations of herself in marriage or parenthood or career, even as events forced alterations. As a young girl she had pictured herself as the wife of a country clergyman, with five children, organizing the lives of the family and the parish, but even as her interest in anthropology was deepening, her husband, Luther, decided to leave the ministry. In 1926 she was told by a doctor that she would never be able to have children and set to work to recast her life. Not long after her return from her first fieldwork in Samoa, she and Luther divorced and she married Reo, whom she had met on the ship returning from the field, intending to establish a childless lifetime partnership in research.

My version of this sequence and how it led to my birth is remembered from what my mother told me when I was eleven, shortly after she and Gregory had also divorced. She took me with her on a trip to Australia and New Zealand where we met Reo, an awkward, gawky man, both shy and abrasive, and she asked me how I would have liked him as a father. I replied that he didn't seem to know much about children. She said that she had married Reo expecting never to have any and had then fallen in love with Gregory as a potential father as well as a scientific collaborator. She did have at least one miscarriage while married to Reo, however, and the outcome was by no means sure with Gregory. Some of what she said, speaking to me about the kind of love that includes the idea of a man, this man, as the father of one's child, may have been an effort to reinforce my sense of the specialness of the man who was in fact my father, although he was the third of three husbands. Just as when one explains to adopted children that they were the chosen ones, this explanation may have been meant to dispel a certain sense of randomness. As a student, when I studied different techniques for drawing kinship diagrams, I experimented with ways of incorporating fictive kin and kinship across ritual bonds when these have been dissolved: What was my relationship, I wondered, to my mother's ex-husbands, those who might have been the parent of her child? What about their spouses and the children they might have? No one else I knew seemed to have such complex networks of people who were not quite kin to imagine relationships with.

Margaret and Gregory returned in 1939 from their last joint fieldwork, in Bali and New Guinea, to be caught up in two uncertain currents, the onset of World War II and my mother's suspected pregnancy. For months it was unsure where my father could most usefully go, where my mother would live, and what the war, preempting all their plans and their imagined futures, would mean. The pattern of life for the next five years was shaped by the war and by their sense of having something essential to contribute to the effort to defeat Hitler. When they discussed it they said that an Axis victory would set science back a hundred years. My mother and I later spent hours in the Vietnam years discussing what it had meant then for intellectuals to be so committed to a military victory—whether indeed one could ever feel such a commitment again after the development of nuclear weapons and whether the new generation could empathize with the past commitment. Thus, although both of them spent some time experimenting to find out how to contribute to the war effort, they lived in a world with one overriding goal, unsure sometimes about how to proceed but clear on the priorities. In later years when Gregory puzzled about the nature of purposiveness, he used to speak of the clarity that a state of war brings as a great relief, of the temptation in any society to resolve ambiguity and hard decisions by turning to warfare.

Gregory had actually traveled to England before I was born, following the instructions of the British consul in New

York that sent him home where he waited disconsolately as it became clear that no one knew how to use his particular set of skills. When the news of my birth arrived, the story goes, he tossed his pipe into the air and over the garden wall, and rushed out to send a cable instructing DO NOT CHRISTEN. My mother always said that this was a replacement for his original text, DO NOT CIRCUMCISE, but given the studied atheism of his family an entirely predictable instruction. The fact that either question occurred to him so late and so abruptly is an indication of his improvisation.

Margaret and Gregory had no household established between their return from the South Pacific and my birth. Instead, my mother brought me from the hospital to the place that was my second and most constant home throughout my childhood, the apartment of her college friend Marie Eichelberger, the place from which I set out for the church on my wedding day. This was a ground-floor apartment in an old house in the Chelsea district of New York City, furnished with the antiques brought from Aunt Marie's childhood in York, Pennsylvania, including a tallboy, a bureau standing some six feet tall with deep, wide drawers, one of which was pressed into service as a crib.

For all the years I have known Aunt Marie, these drawers have always been carefully packed with linens, individually wrapped in tissue paper, tied with thread and labeled, and sewing supplies and fabrics or remnants saved for some future project, as well as mementos of a Pennsylvania Dutch childhood. Aunt Marie is for me the providence represented by that tidy drawer full of little packages, each of which would have been lifted out carefully to store elsewhere, and she is the absolute willingness to dedicate a bureau drawer or a room, an hour or a lifetime to my needs or my mother's.

Aunt Marie was older than the other students when she met Margaret at Barnard and had spent years on a cure for tuberculosis. As a student, although Margaret invited her to be a member of her group, the "Ash Can Cats," she declined, preferring the individual friendship. Later she became a social worker and worked in New York, never marrying, and my mother and I were her family as she participated vicariously in our lives. "You just looked at me across the room," my mother said to her once, "and fell in love with me."

One example of the uncertainties of that period is that, although my parents had agreed to name a daughter Mary Catherine, they had not discussed what I would be called. In the effort not to preempt that decision, my mother carefully called me Sugar until Gregory's arrival in America, varying it through the permutations of diminutives allowed by Balinese morphophonemics: Sugar, Shook, Chook, Chookin, Nyook, ingeniously trying to protect his sense of participation in decision. For years I remained Chook to her rather than Cathy. Friends from that period who had been out of touch would call and say, "How's Sugar?"

I have no record of what I would have been called if I had been a boy, but I know that my father vetoed the name William, the name of his father and grandfather, as "too confusing for librarians," while accepting the name of his maiden aunt, Mary, a pioneer social historian of medieval England who had already given librarians a Mary Bateson to worry about, and that later he named the son of his second marriage for his oldest brother, John. Catherine was for my mother's baby sister, Katherine, born when she was four, whom she was allowed to name and who died within a year. The double-barreled name was also a sort of acrostic establishing a whole set of other references, with the initial *C* replacing the original *K* for my father's mother, Caroline, and the Mary referring to my birthday on the Feast of the Immaculate Conception, which pleased the nuns at the French Hospital, whose support Margaret set out to recruit for feeding on demand instead of on a schedule.

After Gregory arrived from England, Margaret and Gregory set up housekeeping on Ninety-third Street. The pattern of those years affected me a second time when I was making decisions about the care of my daughter, Vanni, and my mother and I looked at films and photographs of my childhood and she advised and reminisced. Margaret wrote that when my father arrived from England, when I was six weeks old, "We let the nurse go and took care of her ourselves for a whole weekend . . ."[1] "A *whole weekend* . . ." Even without italicizing, the phrase bemuses me. All by themselves, they sampled the experience of bathing and changing and caring for a fretful infant and then handed me back to Helen Burroughs, the English nurse Margaret had hired, secure in the sense of having experienced child care.

My mother used to comment that I was doing far too much of the physical care of Vanni in combination with my other work, physical care that I found deeply satisfying. This no doubt corresponded to something I would like to have experienced more fully and richly in my own infancy, ways in which I wanted to make up to Vanni for things I had not had. But for an infant growing up in that era when mother-child relationships were so often scheduled and invaded by technology, I was unusually fortunate, my relationship to my mother firmly established in months of relaxed breast-feeding and many of the chores that go with raising an infant in the practiced and familiar hands of a nurse, the physical care warm and reliable. It seems to me probable that the deficits I was trying to make up came later when that early constellation had been broken by the war.

Margaret always had a multiplicity of rationales for her arrangements. The decision to put me in the care of an English nanny was both an effort at building links and a way of giving her more time and space in which to function. She felt that Gregory, as an English father, would respond best to a child brought to him calm and sweet-smelling and rather briefly in the time-honored manner of an English child, and expected that the echo of English culture in my own behavior would strengthen my relationship to him throughout my life and keep open the option she and Gregory had discussed of living some part of their lives, after the war, in England. In 1947, Margaret took me to England, where we visited Geoffrey Gorer, and was vindicated in this belief when we walked in the gardens at Hampton Court and I bent down to smell the roses, cupping them gently in my hand with the stem between forefinger and middle finger. Geoffrey remarked that I treated the flowers as an English child would have done, an English child instructed by a nanny to "come and see the pretty flowers."

All my mother's arrangements, the choice of a nurse like the choice of a name, had the complicated quality of that kind of lacework that begins with a woven fabric from which threads are drawn and gathered, over which an embroidery is then laid, still without losing the integrity of the original weave. Thus, the choice of my nurse was both a solution to the problems of child care that would permit her her own professional life and an attempt to build a bridge between two cultural traditions and two styles of child rearing. At the same time, it was also a reference back to strategies developed by her own mother, who hired as domestics women with illegitimate children, allowing them to keep the children with them instead of being separated. Nanny, whose husband was long gone from the scene, had an adolescent daughter who lived with us and helped "amuse the baby."

This model, in turn, cropped up in a new variation in my own solution, when my husband and I arranged an apartment, adjoining but separate from our own, which we would rent to a young couple with a child slightly older than our newborn, with a built-in babysitting arrangement. Like my mother, I was trying to combine elements in order to serve many goals. I was trying to build, in a new period, on the situation of the woman who wants to stay home with her child yet feels a need to contribute to household expenses. At the same time I was trying to re-create the form of joint household my mother later shared with the family of Lawrence and Mary Frank, in which the children moved freely between two households, almost like siblings, and the adults had separate and private lives.

When Margaret added a detail to the pattern or made some innovation in the arrangement of life, she was expressing her awareness of how the details of any stable human way of life are linked, interacting in meeting needs and also resonating aesthetically. In a lifetime of rapid change and borrowing of cultural traits from one place to another, a lifetime in which there were continual rents opened in the fabric she worked with, she engaged in constant careful needlework in which repair and elaboration were indistinguishable. When she hung an oval mirror above the marble fireplace in the living room, for example, she thought about what would be reflected in it—two Balinese carvings mirrored within the ornate gold frame—and put a vase from Java at just the right angle to reflect in the mirror as well. She stopped often on the way home from her tower office in the Museum of Natural History to buy gladiolas whose upright forms would reach up into the mirror and echo the Balinese lady opposite with a lotus blossom in her hand.

She was doing the same thing when she hired a nurse, strengthening the echoes in retrospect with a further elaboration of her own motivations and reasoning, so that even the makeshift was eventually stitched into the whole. In the marriage she was the one who set the patterns, for Gregory lacked this fascination with pervasive elaboration. Instead, he was adept at focusing with brilliant clarity on a single point of high patterning, attending to its projections on the surrounding material, but unconcerned by a surround of messiness that was not neatly integrated into the single configuration. Thus, in Bali, he recorded the way in which the hands of men watching cockfights moved in echo of the conflict, but was uninterested in the mass of background detail— uninterested indeed in photographing the fighting cocks themselves. His life was full of loose ends and unstitched edges, while for Margaret each thread became an occasion for embroidery.

Margaret planned in patterns and watched for and accentuated those she found, giving to many decisions that others would take as a matter of course a special quality of purposefulness. Even those things that may have been accidental or simply assumed were afterward treated as a part of the pattern. It is not surprising that the things that I know and have always known about my own early childhood turn up repeatedly in her writings, stitched and restitched into the myth of herself and the myth she gave me of myself. But the elaboration of detail and the wish to specify it exactly were rooted in precise and elaborate fieldwork, ethnographic study of eight different peoples, each handling the details of life differently in the certainty that their way, the way it had always been done, was the right way. Once, after the death of her sister Priscilla, Margaret asked a florist to put together some distinctive and clearly visualized arrangement and he rebuked her, explaining that it simply was "not done" that way. She stamped her foot and said, "I've organized funerals on four continents, and I'm going to have the flowers the way I want them."

Having the long-hoped-for baby was also a chance to choose purposefully among many alternatives she had seen, or rather to make selective variations in the customary that she felt would be improvements, and doing this carried with it the obligation to observe. At one period, they planned to set up floodlights in every room of the apartment so that it would be possible to record immediately any interesting piece of infant behavior. My early memories of my father always include the Leica that hung around his neck. Since Gregory was away when I was born, my mother arranged for somebody else to film my birth, a procedure that was almost unheard-of in those days. Margaret believed that an infant, in the hour after birth, unless she is heavily affected by anesthetics given to the mother or battered by a particularly difficult passage, is more clearly herself than she will be again for days or months as the environment makes an increasing mark, so that these moments were critical to record.

Recorded they were, in detail, with a series of neurological tests and manipulations that are disturbing to see on the film today, with our growing sense of the importance of tenderness in the delivery room. But seeing that film before my own infant was born was probably helpful to me in making me less fearful of the fragility of the newborn, as the infant on the screen—myself—is poked and tickled, bent and dangled, howling and finally exhausted. My friends ask whether it makes me angry to see that, and I respond, no, here I am, I'm okay.

When Margaret planned for my care and feeding, she set out to combine the generosity of most primitive mothers, who nurse their infants when they cry and remain with them constantly, with the resource of civilization, the clock, and this too meant recording. She would record the hours at which I demanded feeding and then, by analyzing these times, construct a schedule from the order immanent in my own body's rhythms which would make the process predictable enough so she could schedule her classes and meetings and know when she should be home to feed me. I have the notebook that records these feeding times and other observations, as I also have the notebook in which my grandmother recorded her observations of the infant Margaret and the one in which I recorded my observations of my daughter, Vanni, while I went off between feedings to teach a seminar and analyze films of the interactions of other mothers with their children.

Margaret's ideas influenced the rearing of countless children, not only through her own writings but through the writings of Benjamin Spock, who was my pediatrician and for whom I was the first breast-fed and "self-demand" baby he had encountered.[2] If the weight of early experience is as great as we believe it to be, I belong to a generation that is chronologically some five years younger than I, psychologically one of the postwar babies although I was born in 1939. What Spock finally wrote about "self-demand" after the war, however, was not quite the same as the method Margaret developed, for Spock advised mothers not to enforce a schedule immediately but to wait and shift infants gradually into the classic feeding times, rather than assuming they would develop and retain individual rhythms.

Spock was blessedly relaxed about letting my mother do as she wanted, abandoning the fixed schedules that were regarded as essential to health, but he seems to have been only partly aware of the innovation taking place in front of his eyes, for he wrote later that the first experiment in "self-demand feeding" took place in 1942,[3] an example of the limited willingness of physicians to learn from patients. In 1969 when Vanni was born and I rethought these questions, it seemed to me that Spock's approach corresponded to the false permissiveness of campus administrations all across the country, urging that students be given a degree of freedom *until they settled down*, rather than really listening and taking seriously what young people were saying, the emergent clarity of their own visions, like the order immanent in an infant's bodily processes.

The photographs and films that were made of my infancy do not really belong to me, are not private records. Because of the war, the series was broken and has less value than Margaret hoped it would, something I am grateful for, since it meant that her discussions of my childhood stayed within an anecdotal framework. Still, when I wanted as a college student to discard a great stack of my childhood paintings, my mother told me I had no right to do so—that I had probably had the best-documented childhood in the United States. From the recording of a unique case of innovation something might be learned, and so it was subject to her ethical commitment to collecting and sharing knowledge. Still, she was not experimenting in the usual sense of the term, for she could have made the decisions she made only in the conviction, already arrived at, that her innovations would be improvements, that her knowledge of human cultural diversity could be combined with the possibilities of technology and modern medicine to improve conditions at the beginning of life. The process of inquiry, involving the life of a child, could have been pursued only in a context of advocacy, and advocacy, for Margaret, was never far behind.

Anthropologists sometimes speak of the field as their laboratory, but in general our knowledge is based on observation rather than on manipulation. Where we act for change it is to achieve goals seen as valuable rather than to generate data. Usually, our experiments are those arranged by history and most of our variables are embedded in the flux of human life and cannot be isolated, having neither beginning nor end but unfolding over time. Traditionally anthropologists have not been able to work with randomly selected populations and matched controls; their data have come from the observation of unique individuals and could be compared from one to another only because the place of each was known. Each life history, and the record of each community with its own distinctive and interlocking patterns of adaptation, is valuable and to be recorded, a unique experiment.

Margaret was experimenting as a painter or a sculptor experiments, innovating against the background of tradition and previous work in response to a changing imagination, following an emergent certainty of the place for a brush-stroke or a patch of color, seeking and expressing not curiosity alone but the discovery of conviction. She retained the humanistic awareness that the creation will be unique and the scientific belief that the process is not finally a private one. She was recording as anthropologists always record, in the hope that whatever the outcome, when the unique instance is fully placed and described it will be an addition to our understanding of the human condition.

I have wondered sometimes about her assurance, since she was doing things that were widely believed to be wrong or unhealthy for infants, calmly planning how to bully doctors and vamp nurses into allowing all sorts of irregularities at a time when most women find themselves easily bullied by those who represent medical authority. When Vanni was born, I was enriched by my mother's confidence, becoming able in my turn to reject the kinds of advice that undermine breast-feeding and invade the intimacy of mother and child. It was splendid, for instance, to have Margaret robustly declare that it was rubbish that I should never nurse Vanni in bed for fear of dozing off and suffocating her, as all the nurses insisted. All around the world mothers and infants sleep side by side and the danger of suffocation arises mainly when mothers are drunk or sick, not under normal circumstances—it is the American habit of leaving an infant alone in a crib in a separate room that is at odds with the normal range of human behavior.

Some of her assurance came from having seen alternative ways of doing things, healthy mothers and infants thriving within a variety of different patterns. This range gave her a sense, behind the diversity, of what was essential, different from the assurance of those who take tradition as their only base. She had been raised in a context of educational experimentation and she was working in a time of newly vivid awareness of the damage that could be done by Western and "modern" forms of child care, of the burdens of neurosis carried by members of her own generation. Indeed, she selected Spock as a pediatrician because he had been psychoanalyzed.

Nevertheless, she would have known that there were

risks in her innovations. Details of infant care are helpful or harmful depending on the way they fit into the rest of experience. She often told stories of incongruous or incomplete cultural borrowing—a change in patterns of food preparation that leaves out an essential nutrient, or the way that many Western mothers, imitating other methods of child rearing, carry their infants on the right hip, making themselves unable to work and depriving the infant of their heartbeat. There is always the risk of crippling some basic biological capacity, as has been demonstrated in experiments with distortions of the infant experiences of monkeys or birds, which then grow up unable to mate or unable to care for their young. The more important risk was that some changed constellation of infant experience would set me at odds with my own society in subtle ways or leave me unable to adjust to later challenges, to fall in love, to care for a child, to function in relation to contemporaries whose early experiences were profoundly different from my own. I was inordinately proud, as a child, of having been a "self-demand baby," but in other periods I puzzled about whether I was different and whether there was something in my childhood that made it seem so difficult for me as an adolescent to blend in and be like everyone else.

It seems to me in retrospect that Margaret's willingness to make innovations came out of a certainty of her own love, a sense that she had been loved and could trust herself to love in turn, with a continuity of spontaneous feeling even where she was introducing variation. She was prepared to take responsibility because she did not suspect herself of buried ambivalence either toward me or toward her own parents. Indeed, in a life lived in an era of introspection and self-doubt, her conviction of undivided motives was distinctive, an innocence that leaves me sometimes skeptical and sometimes awed. Just as all of her commentaries about American culture and suggestions for alternative arrangements must be read against her general affirmation of the American tradition, so her sense of choosing her own style in child rearing was secured by her appreciation of her own childhood and her desire for motherhood, for she believed that these would protect her from destructive choices. She drew an immense freedom from her conviction that she had no inherent temptation to destruction and that the arrangements that best served her professional life, given her ingenuity, were in no inherent conflict with my welfare. Over the years this attitude was contrasted with cultural styles that depend upon a suspicion of one's own cruel or evil impulses, as English children are taught to be kind to animals because of the temptation of cruelty.

My mother and I used to discuss sometimes which parts of her approach to child rearing seemed right in retrospect and which needed to be thought through again. Both of us felt comfortable about self-demand feeding and I followed her version with Vanni, but there were areas where more or less pattern seemed to be necessary. I argued with her, for instance, that pleasant as it is to have an unroutinized body in a society where others are dogged with concerns about "regularity," a little more routine would be useful—some points should be fixed in the day, if only so that one could, if necessary, remember to take medication on schedule. Thus, she never insisted that I brush my teeth because she had found the process unpleasant and painful as a child, so this was a routine I had to establish on my own, that required attention at a stage when most of my contemporaries did it automatically—but then, I don't suffer when some hitch in arrangements separates me from my toothbrush for twenty-four hours. Another area that we debated was sleep and how to arrive at a balance between a child's sense of her own needs and the patterns of the day. I resisted sleep all through my childhood, giving Vanni in turn more leeway to find her own rhythms, and I teased Margaret at her inconsistency in not applying the concept of self-demand to sleep as well.

Margaret's childhood experience had made her different from others, but generally in ways that she found rewarding rather than alienating, and this was something she wanted to pass on. She spoke, for instance, of having been brought up by parents who were genuinely not racist and thus of not sharing in the residual guilt that others use to mobilize commitment. Similarly, I sometimes wish I could share more in the feminist anger of my contemporaries; but even though anger and guilt are useful in many situations, they carry great costs. There is no way finally that I can evaluate the extent of my own difference or how much this is related to infant experience. For it seems to me that all of us share to some degree in the experience of unintelligibility, sometimes feeling less than we might have been, or uncomfortable in our own skins and alien from those around us. I have always tended to look to the special circumstances of my childhood whenever I felt unhappy or lacking in confidence, and yet it is not reasonable to attribute a degree of estrangement that is part of the general human condition to a particular idiosyncratic experience. There is no form of human child rearing that does not leave an occasional residue of fear and yearning; these are part of a common inheritance matched in the wider culture by at least partial forms of solace, with as many forms of psychotherapy as there are forms of ritual and belief. In this country, too, difference is part of what we share.

The innovations that Margaret made as a parent were actually greater than they now seem because so many have since been incorporated in the patterns of the society. They were balanced, however, by patterns of conservation, by the delight Margaret took in tradition and the desire to preserve complicated kinds of intelligibility. No one with an English nanny can grow up without a sense of continuity. Over time it turned out that the most important characteristic for me to have—adaptability to different kinds of people and situations—was also what Margaret would come to care most about providing for a generation that was to grow up in the midst of change.

As I write, I am struck by how little my father comes into the discussion of the details of my infancy. Usually he was behind the camera, not in the picture. I know he found great pleasure in his children as soon as we were old enough to begin to talk and play, but I suspect that his relationship with me as a tiny infant was a gingerly one, a carefully monitored delight. My mother emphasized the role of fathers in relation to infants less than has become fashionable in recent years. She was leery of the father's presence in the delivery room, feeling that this role belonged to another woman, one who had experience of childbirth, a grandmother or perhaps, as among the New Guinea Arapesh, the woman who has most recently given birth.

When Vanni was born, even as I was taking notes about the way in which my own reactions to the newborn were biologically activated, Margaret argued that the emotions to be activated in a father were those of caring for and protecting the mother-infant pair, not the individual infant. At the same time she argued that if a diffident new father was allowed to put off taking the fragile-seeming newborn in his arms, it might be six months before the inhibition was overcome. She talked of the sweet and inoffensive smell of infant feces, so long as they are receiving breast milk only, but she did not seem to expect that a husband or father would be tolerant of this or any other smells, and advised me to give Vanni vitamin drops instead of orange juice so that when she spat up there would be no sour odor. This was Margaret, however. Gregory was basically indifferent to wet spots on his knees, to stains and spills, from children or animals, and indeed to the stings or scratches he got from handling wild creatures. His tenderness, whether to children or to animals, always contained an element of the naturalist's care, tolerant and admiring of living grace.

The first few images of the transition from infancy to childhood, fragmentary glimpses only: walks with Margaret and Gregory on which I was held securely on a leash, so that I could safely range farther than a child held by the hand; trips to the park when my father would push the swing high enough to run under it; Nanny's horror when I found a longicorn beetle which I remember as a "laundry bug"; sitting with my parents as they drank coffee in the morning and made a pair of courting "light birds" dart across the walls and ceilings by reflecting light from the bowls of silver spoons.

The continuity of that household, the matter-of-fact and slightly astringent affection of Nanny and the shared affection of parents still happy in their marriage and delighted at having a child, continued until I was two. The modern appraisal of the importance of early infancy was just beginning to be widespread in those days, but there is no question that Margaret and Gregory were consciously trying to establish in me a base of trust and self-confidence. On two occasions, once when I was eighteen months old and once when I was two, they worked on plans, never realized, for teaching films that were to convey an understanding of what trust and confidence mean in a child, and I was to be that child, to exemplify that trust and confidence. Margaret wrote in *Blackberry Winter*, "How much was temperament? How much was felicitous accident? How much could be attributed to upbringing? We may never know. Certainly all a mother and father can claim credit for is that they have not marred a child in any recognizable way. For the total adult-child situation could be fully understood only if one also had the child's own interpretation of the parts that adults played in its life."[4]

This book cannot be the child's interpretation, for that child is now an adult, and what I write about that period is a reconstruction. Nevertheless, it has seemed to me, when I pass through difficult periods and try to wrestle through to a better understanding of myself, that behind the veils of perplexity or despondency I find resources of faith and strength, a foundation that must have been built in those two years.

When my daughter was an infant, I took Margaret's advice not to worry greatly about persuading her to sleep through the night, and enjoyed my nighttime interludes of nursing in the dark and silent house where no telephone rang, sitting in a rocking chair my mother gave me and wrapped in a shawl sent by my husband's mother. I had pursued a more immediate intimacy with Vanni than my mother had wanted to, indeed, been allowed, handing over less to other caretakers and nursing longer. I was also more relaxed about the resilience of an infant, and I watched with amusement when Margaret visited and carefully washed her hands and changed out of street clothes before she approached the baby, still knowing and full of good advice but slightly tentative.

I got many of my ideas about child rearing from my mother but I also read "Doctor Spock" in my turn. Gregory and his third wife, Lois, had a daughter a year and a half before Vanni was born, and many of my ideas came through Lois's contact with women who were thinking about home birth and breast-feeding, in a new interweaving of the generations. In Vanni's infancy, Margaret again brought photographers, this time to take illustrations for her article "On Being a Grandmother,"[5] and for other films and writings. The discussions of child development went on into a new generation as I contributed my awareness of the development of communication. But the real clue to understanding the quality of the infancy my mother gave me was my own deep pleasure in mothering, and the sense I had that the innovations I wanted to make in my turn were essentially affirmations.

III
A Household Common
and Uncommon

The pattern Margaret and Gregory established in their apartment on Ninety-third Street, with Nanny and her daughter to help in looking after me, was varied by summers in the country when the entire household went to stay in rented quarters near Holderness, New Hampshire. The Bateson family, like other families of social scientists over the years, was drawn there by Lawrence Frank, who was emerging as a central figure in setting the pattern for interdisciplinary cooperation in the social sciences. Finally, in 1942, when I was two, we moved into the Frank house in New York and the households were merged, winter and summer, "for the duration."

That decision was made on December 7, 1941, the day before my second birthday, Margaret and Larry Frank were together at a conference, one of a series Larry had organized, and when the news came of the raid on Pearl Harbor, they turned to plans for mobilizing themselves for the war, for the Japanese attack meant American involvement in both Europe and Asia. It meant that this would be a world war. Ruth

Benedict, Margaret's friend and anthropological colleague from Columbia, proposed that Margaret go to Washington, where she would work with the Committee on Food Habits, and Larry proposed that the Batesons move into his house at 72 Perry Street in Greenwich Village. Larry was a widower with a new and much younger wife, Mary, an infant son, Colin, and the five older children of his previous marriages.

My own coherent memories of childhood begin at Perry Street, a complex household with much coming and going, a huge table spread with great meals for which Mary baked and baked. It was a five-story brownstone house, the bottom two floors set aside for the Batesons, the Franks on the upper three. During the war the downstairs was vacant for long periods and my life was upstairs with the Franks, Uncle Larry and Aunt Mary, while my parents came and went, longed for and welcomed transients. Thus, I did not grow up in a nuclear family or as an only child, but as a member of a flexible and welcoming extended family, full of children of all ages, in which five or six pairs of hands could be mobilized to shell peas or dry dishes. My mother paid for servants of various kinds over the years and Aunt Mary provided a guiding and discriminating maternal spirit. She was beautiful, Irish, and very young to take on all that household, and she danced jigs for Colin and me in the living room, balancing a broomstick on her finger. Friends expected my mother to be jealous, asking, "Don't you mind handing over the care of your child to a younger and more beautiful woman?," but Margaret argued, then and later, that jealousy was culturally produced, an emotion she did not feel.

Margaret believed that it was not only possible but preferable that children feel a part of several households and have several caretakers. In this way, she believed, it would be

possible to avoid the tightness of bonding to a single caretaker that so often provides the ground of an entire neurotic
system. In Samoa, she had described a number of instances
where children simply moved into a different household.
Thus, the invitation to live at Perry Street must have come
not only as an accommodation to the necessities of wartime
but as an opportunity to function more freely herself and to
construct the kind of household she felt was best.

The wartime arrangement with the Franks, which continued after the war when we remained in the same house
although the households were more clearly separated, was
always described by us to each other as a utopia, and the
summers in which the households were merged at Larry's
house, Cloverly, were always recalled as equally ideal.
Indeed, the same rosy light that pervades my mother's picture
of Samoa surrounds the picture of those New Hampshire
summers. Instead of the palm trees and the surf, the image is
of evergreens with their own murmuring and the lake
beyond, a place where there was no anger and no grief,
where each child was cared for by enough adults so that
there need be no jealousy, where the garden bloomed and the
evenings ended in song. It takes a major effort of will to
remember that there was any stress at all.

In World War II, when fathers went off to war and mothers to work in industry or agriculture to support the war
effort, these separations were conceived as sacrifices. The
secret reminiscences of a temporary freedom, male cama-
raderie for the men, self-reliance and competence for the
women, slightly shameful pleasures drawn from a world in
travail, are hidden, and the separations of the war years are
depicted as times of deprivation and endurance. But I grew
up being told that although my mother felt she had made a

sacrifice and would have enjoyed being with me more during those years, I lost nothing. Whatever it felt like to have both parents progressively withdraw, beginning when I was just over two years old, and then be absent for increasing periods of time, these memories are almost entirely lost, as is the memory of the stress created by introducing an additional child into an already complex household where a younger infant had a right to those years as the center of concern and delight. A sense of aloneness, a sense of an occasional frighteningly inimical response to midnight demands for comfort—these must be dredged from the deepest part of dreams and then combined with the adult knowledge that any caretaker would be exasperated and burdened under those circumstances. I was rich beyond other children and I had the delight of a playmate who was like a brother, Colin, and yet there were all those partings. There were all those beloved people, yet often the people I wanted most were absent.

Once the decisions were made, transitions followed rapidly. Between Pearl Harbor and the summer, my mother started commuting to Washington. In the summer of 1942 we were all at Cloverly, but in the fall Nanny left, following her own preference for very young children. In 1943 my father started working in Washington and a little later that year, when I was three, my mother went to England to lecture on the differences between English and American character, staying longer than was expected. Later she told me that when she returned I had snubbed her as children will when they feel betrayed. No one had told me when her trip was extended, not expecting me to be aware of dates or sensitive to changes in a total period of several months, but I had calibrated my expectation of her coming by the seasons. She was to return during the summer, before nursery school started, but the leaves fell and school opened and she was still missing. She developed then the idea that I could handle absences and disappointments but not violations in communication, that it would have been all right to extend her absence if I had been kept informed. Her rule from very early on was to provide for either continuity of place or of person—never to leave me with a strange person in a strange place—and although she never herself owned a house, the houses and apartments of my childhood, even the empty downstairs rooms, with their tall white-curtained windows, remained places of comfort.

For two years or so both parents came and went, from Washington or farther afield. I remember the feeling of playing house when I went down to breakfast with them when both were there for the weekend. My mother bought a fine china demitasse set so I could serve afternoon tea to my English father when he was home for the day. I remember being demure and proper to see a father who was often absent, and then holding on to his raincoat and screaming when he had to leave. In those days my mother wore coarse tweed suits that were so scratchy that when I sat on her lap I brought a handkerchief or a little pillow to rest my cheek against.

In 1944, having failed to find a place in the British war effort, my father went to the Pacific with the Office of Strategic Services. I was luckier than many children, for I was four by then, old enough to remember his departure "for war." Before he left, Margaret had him write dozens of postcards addressed to me. When weeks would go by without a word and someone would ask, "Any word from Gregory?" she would say, "I haven't heard anything," and then I would say,

"But I have." In Burma they had a house mongoose, and Gregory sent home photographs of himself with the mongoose perched on his shoulder. He moved around a good deal, designing various kinds of propaganda and describing social structures in exotic contexts where it was hard for Americans to know even whom to talk to.

He used to tell of how he and a friend, Jim Mysbergh, conceived the notion that Japanese morale would be shattered if the cremated ashes of Japanese soldiers, appropriately packaged and with appropriate accompanying texts and prayers, were parachuted into Japan. The gap between theory and practice was wide, however, and when by accident a drowned Japanese airman was washed up on the Pacific island where they were stationed, the logistics defeated them. An attempt at cremation was interrupted by wind and left them running up and down the beach trying to gather the right ashes, and months passed before the appropriate expertise could be found for the packaging. That done, the U.S. Air Force flatly refused to be involved in what must have seemed a ghoulish operation. The ashes of "Poor Yorick" were carried around until the war's end, remaining for Gregory an example of an ingenuous misplaced concreteness: Why human ashes? Why *Japanese* ashes? And why tell the airman what the package contained?

According to another story, they set out, in Burma, to dye the Irrawaddy River red, in fulfillment of some apocalyptic local prophecy—but did they really stand, knee-deep in the river, pouring in dye themselves as it was diluted and carried away almost without effect? Or is that just a child's imagining of some more efficient military effort? Ten years later, when my father was applying for American citizenship, having lived all those years in America as a British subject, one of the points of doubt that were raised was his years in an intelligence service of a foreign power—the foreign power being Britain's ally, the United States.

The war in Europe ended, but he did not return. I remember understanding for the first time that there were really two wars, that Europe and Asia were different places. Otherwise the whole process of the war was very remote, full of unfamiliar names. I have a sense of the commitment of the Cloverly household to growing and canning fruits and vegetables, the careful counting of ration coupons, the saving of tin cans. At noon the household would gather around a bulky radio at one end of the dining-room table and listen to the news. Roosevelt died and we saw strangers weeping in the street. In 1945 the first atomic bomb was exploded over Hiroshima and as children we knew something of it, for when we were asked to draw pictures of the bomb at nursery school, I drew a series, from different points of view, including one from the ground before the explosion: "This is the way it looks before it goes off, but in a minute there won't be anyone alive here to see anything."

And even so, it took a long time in that year of parades and triumphs before Gregory was back in New York, and no one could predict exactly when he would arrive. At last one day my mother brought him to my school to find me. I was outside my kindergarten classroom getting water for the plants, and I ran down the hall to her, boasting of my task. So great was the difference in height between them, and so remote was my father's height for a five-year-old, that I remember only a shadowy pair of trousers, a raincoat, "Chook, do you see who's here?," and my eyes traveling upward until I screamed in delight.

The Christmas after Gregory returned from the Pacific,

the first Christmas of peace, there were still shortages but at the same time there was an efflorescence of imagination and ingenuity. Margaret made the invention of stuffing and roasting frying chickens because larger ones were unobtainable. New kinds of Christmas ornaments appeared in the shops, many made with pieces of mirrors, many of them angels. Gregory anchored our eight-foot tree to the elaborate moldings of the high ceiling and the window frame, and then spent hours trying to discover which of the colored light bulbs, all wired in series circuits, were burned out. In front of each bulb he spread a filter of angel's hair, which came on the market for the first time that year, presumably a by-product of wartime development, and he set a wooden angel in front of the star at the top of the tree. She is white-painted wood— we still have her—and she has traveled around the world with my husband and me along with the other carefully treasured postwar novelties and with ornaments going back to my mother's childhood—a china doll with a white fur body, a robin redbreast, Viennese miniatures of musical instruments. After hours of work on that tree, having made sure that a white light came out at the feet of the angel, my father carefully and precisely broke the side of a silver glass ball to set the hemisphere over the bulb, reflecting light up into her face. Two Christmases he did that? Three?

We lived together for only about a year after Gregory's return. During that winter, there were many loose ends from the war which required both of them to travel, so that I have very little sense of the three of us living on Perry Street and I must have gone on for a time sleeping upstairs in the Franks' big playroom. My real memories of living together come from the summer of 1946, when my parents rented a house about half a mile from the Franks, called Briarfield, and we spent the summer together there. I was six years old.

My father was appalled to discover that I could barely read. Sitting on the screen porch as I sounded out the words of *The King of the Golden River*, he set out to teach me. One day in exasperation he pulled out a copy of *Naven*, his own book about New Guinea, and pointed to his name on the front, and I was totally unable to make out the Gregory until I saw the Bateson. "What *do* they teach these children at school?" is an archetypal English paternal remark, and one that he kept on making through the years, expressing skepticism about all the options of education in America. In the end he let his second daughter, my half sister, Nora, drop out of elementary school completely for several years.

During that same summer, he set out to teach me a little natural history, since he had spent his boyhood collecting insects and butterflies. He bought a microscope and a telescope and we spent hours looking through both. We went off together and collected all sorts of creatures from the lake and the swampy area on the other side, taking them home in jars. We carried out such simple experiments as standing various plants and vegetables in ink or dropping light and heavy objects from the upstairs windows. It seems to me that he put a lot of thought and imagination into devising the pathways of discovery down which he wanted to lead me, even as I was rediscovering the pleasure of being with him.

Both my parents loved reading aloud and being read to, especially poetry. That summer resonates with *The Rime of the Ancient Mariner* and Walter de la Mare's description of his own boyhood discovery of poetry. The tangles of dream and nightmare in the poems stay in memory beside the pictures of that brief interlude as a family, recalled as idyllic and filled with sunshine.

Briarfield is an old house with tall brick chimneys and great evergreens behind it. There was a rather formal parlor with stiff Victorian furniture and a curtain across a double doorway at one end, like a theater. A project developed to perform the scenes in Shakespeare that seemed to prefigure contemporary psychiatric insights, still fresh and exciting to other social scientists. Uncle Larry and my mother did not act, but directed. During the performance, however, they conducted a dialogue commenting on the whole enterprise. As I saw it, they *sat* and they *talked*, as they did through a dozen subsequent summers, getting up at six and rocking on the long Cloverly porch for hours on end. Gregory played Hamlet to Aunt Mary's blithe Ophelia, and Larry's daughter Barbie, as Gertrude, seemed years older than her stepmother. The primary emphasis in the choice of the scenes from *Hamlet* was Oedipal, but we had Ophelia's mad scene too, and much of the conversation that summer must have concerned schizophrenia, since the work my parents did in Bali had been supported by a fund for the study of dementia praecox. Indeed, "There's method in his madness" might be taken as the motto of all Gregory's later schizophrenia research. We had the sleepwalking scene from the *Macbeth*, with the doctor played by another of the older Frank children, Alan, who was planning to go to medical school and eventually became a psychiatrist. Then Alan and his sweetheart played the balcony scene, perhaps because it fit in with what was being said by semanticists, perhaps because of a sense that falling in love is a kind of madness, a part of the poignancy of adolescence, perhaps simply to weave in a pair of young lovers— there was always a pair of lovers at Cloverly. We had scenes from *Twelfth Night* as well, but curiously we had neither *The Tempest* nor *Lear*. I think the world was very young.

Rehearsals and costume making went on all around us for weeks. Colin Frank and I were too young to participate but we watched in fascination and learned how to make stiff Elizabethan ruffs for a rubber dog we played with while more elaborate ones were sewn for Malvolio. Margaret would become exasperated with Gregory during rehearsals as he altered Shakespeare's wording, protesting that the meaning was preserved. On the day of the performance, when there was a delay, I went in with my friend Martha Ullman, dressed as gypsies and given the task of ushering, and we recited passages to the restless audience.

In *Blackberry Winter*, Margaret described the amateur theatricals she organized as a child, but also said of her episodic experiences in different schools, "In all the schools I had attended so far I felt as if I were in some way taking part in a theatrical performance in which I had a role to play and had to find actors to take the other parts,"[1] including a best friend, a boy to fall in love with, and a teacher to attach herself to. Cloverly was a stage, and there and elsewhere Larry Frank achieved a special idiosyncratic kind of greatness as a creator of complex groups that would both work and play together. The assembling of groups where insights would complement each other became a theme in both my parents' lives.

Larry liked to work with his hands, spending several hours a day working in the garden. He was always interested in new technologies and machinery. My sense of him with the household or with any group was as a gardener or husbandman, and his interest in health was pervasive. One of the first social scientists to become interested in nutrition, he made up a set of vitamins each day for every member of the family. The colorful pills appeared in little colored glass dishes by each place at the breakfast table, as three generations came and went. One day in the summer of 1946, I am told, I cut my bare foot walking along the road between Cloverly and Briarfield. Whoever was with me—one of the adolescents— judging we were about halfway between, asked which way I would like to go, and after some thought I said, "Daddy is good with nature, but Uncle Larry is better with wounds."

We also had a cocktail party that summer. It was part of the immediate postwar style, and very different from the kind of parties the Franks gave in New Hampshire, big square dances and picnics and corn roasts that combined all the generations. My mother and I spent a day preparing canapés and making animals out of fruits and vegetables and toothpicks to decorate each platter, like the elaborate offerings made of food that the Balinese prepare for their gods. We had no refrigerator at Briarfield and had to have ice delivered for our icebox. When I try to think of my mother making things from materials other than words, I think of food and knitting. Yet I know that with the young girls in Samoa she learned to weave baskets and grass skirts, that ethnography required all sorts of manual skills, and in *Blackberry Winter* she makes it clear that skill in various arts and craft activities had an important place in family ideas about education. She had few such skills in my childhood. She had great difficulty threading needles, and in later years Aunt Marie always made sure there was a supply of ready-threaded needles in the sewing box for emergency repairs.

Margaret was very selective in her modes of participation. She engaged in almost no physical activity other than walking a little (but not hiking) and floating dreamily in the water with a balsa board or a tire, although she didn't swim. (Later on, when it became urgent that she do some little physical exercise, she was able to take an interest in exercise only with a complicated rationale and associated mental activity.) Thus, she was an onlooker as others square-danced or played tennis or croquet, a passenger in the canoe or row-boat going to picnic on the island.

There was a whole series of activities that she had decided she could not excel in and had withdrawn from, and although she had taken drawing lessons as a child I could never get her to draw for me. There were other things that she said bored her, including all board and card games, and she had strategically decided not to be able to play bridge as a way of limiting participation in the social life of colonial administrators in the field. In fact, she abstained from all games that were not open-ended to the imagination, but she was wonderful at contriving games that would have that open-ended quality and would lend themselves to different levels of participation. Photography was pervasive but left to my father.

Thus it was that as a child I thought of Margaret and Larry Frank and Ruth Benedict as belonging to the same generation, as choreographers, of Aunt Mary and my father as a different and younger generation who still danced, even though my mother and my father had a scant difference of three years between them. My father used to go places with me, the woods and the swamp, while my mother was normally sitting down. Once in that period I said to some adult who repeated it back, "My poor Mummy, she's so little and so old." At the same time, although I had no inkling of trouble, the time was approaching when Gregory felt that to move ahead with his own life, he would have to leave Margaret, bringing into his conflicts with her a store of hoarded hostility to his mother. He brought to the marriage the experience of being the youngest child of his own family, whereas my mother went through life acting and feeling like a responsible oldest child.

What were the games we played? Twenty questions I associate with my father, a game in which the binary discriminations were key. We did not start from "animal, mineral, or vegetable" (a very unsatisfactory division), but from "abstract or concrete." For the children, the answers were sometimes concrete, but for the adults they were usually abstract: tomorrow morning, the difference between truth and beauty, the superego. Under these circumstances, there was little point in counting questions, one simply competed against a standard of logical elegance in partitioning the universe of possibilities, and in the intervals explained to the children the mathematics that made this possible.

Another splendid game was called conversation. Two people would be sent out of the room and each would secretly be given a sentence he or she had to say in the course of a conversation. The rules required, however, a ceremonial introduction in relation to some third theme, followed by conversation that was logical, courteous, and intrinsically interesting. I remember once two psychoanalysts were pitted against each other, one of them instructed to ask, "Have you ever seen an Australian tag wrestling team?" and the other, "Don't you think men knit better than women?" Introduced to each other as members of the same profession, as colleagues, one launched off on a discussion of what it would mean to have alternating analysts, while the other edged toward a discussion of an analyst knitting during sessions.

The Cloverly game par excellence was charades. Our

charades were full-blown performances, plays with dialogue
and costumes and props. Each charade was a polysyllabic
word, acted out syllable by syllable and then as a whole, but
the rules required that the series of scenes hang together in a
single narrative, with a continuing group of characters. The
adults kept themselves busy devising complicated ideas while
the children and adolescents did most of the acting. Once
Colin and I played Adam and Eve, myself at six in the buff
but Colin at five preferring to wear a bathing suit. Playing in
this way, we would rarely manage three charades in an
evening. As teachers, Gregory and Margaret both liked to
devise examinations with this same kind of complex integra-
tion, several questions to be addressed in a single essay.

My father, having learned to drive during the war, got his
driving license that summer, but Mary did almost all the driving and Larry and my mother never learned to drive. The
Franks had a Model A Ford called the Bluey, with the true
classical *oogah* horn, and with the top down it expanded like
an illustration for *Cheaper by the Dozen* (a book the Cloverly
family read with great pleasure and some recognition). My
mother had once started learning to drive but the process was
interrupted by a minor accident and never taken up again.

Larry's first major interdisciplinary conference had been
held in an inn in Hanover in 1934. In subsequent years he
made Cloverly the center of a community of social scientists
who rented or built nearby. After Margaret and Gregory separated, I spent months each summer at Cloverly while my
mother came and went, spending her days in conversation
with Larry or typing in a small cabin by the lakeshore just
remote enough from the landing to allow undisturbed work.
The pattern of life in those presuburban days, if one could
afford it, was to live in the city and then go away in the sum-

mer from the heat and the polio epidemics. Every year there
was a major hegira, with an eight-hour journey by train, broken by a change in Boston.

There were many famous names in the company that
gathered at Cloverly in those years, but it was only when I
read their work in college that I began to have a sense of who
they were, and the things I remember best are not the things
they became famous for. Gardner and Lois Murphy, the psychologists, built a house down the road after renting in the
area, and we all gathered to see the ground cleared and to
hike far up the hill behind the house to see an ancient "wishing tree." Since they had no lake frontage, they would slip
through the Cloverly garden as they went down to the landing to take their canoe out in the evenings. Norbert Wiener,
the father of cybernetics, used to stop by, smoking smelly cig-
ars and pouring out his latest idea to Larry or Margaret, not
much interested in listening to their responses. Ruth Benedict
used to come there, as did Erik and Joan Erikson. Many of
them had adolescent sons and daughters a few years older
than Colin and me, some of them friendly to little children
and others intolerant. Many of the older generation are dead
now. Of Robert and Helen Lynd, the authors of *Middletown*,
whose teen-aged son Staughton was my knight errant when I
was six, I remember best a song about an animals' picnic that
Bob used to sing at the very end of long evenings around a
campfire, with its mournful chorus that fades away after the
animals die one by one.

> *Come jine the huckleberry picnic,*
> *That's gonna take place today.*
> *I'ze on a committee to invite you all,*
> *But I ain't got long to stay.*

IV
"Daddy, Teach Me Something"

After the Briarfield summer, Gregory withdrew from the Perry Street household, first to a borrowed apartment and then to Staten Island. He used to pick me up at Aunt Marie's apartment and take me out to the island by ferry. Margaret made the effort to avoid any situation in which I might see conflict between them or get a sense of the decay of the marriage or the love affair he was having with a dancer. Indeed, for a long time she made sure that Aunt Marie was unaware of the deteriorating relationship so that I would not sense it from her. She used to talk about an occasion when her voice broke and I darted across the room to find her looking out through one of the tall windows, covering up the fact that she was at the edge of tears. Together Gregory and I took long walks along the shore. I don't know whether my memory of bleakness comes from his depression or whether I am responding to the barrenness of those winter beaches, but I remember them as gray and devoid of life.

It was apparently Margaret who persuaded Gregory to

go to the Jungian analyst and common friend, Elisabeth Hellersberg, and, after a certain amount of three-way communication, she in turn persuaded Margaret to meet with another analyst for support as long as the situation continued, but Margaret's efforts to accommodate the transitions in Gregory's life seemed to him a continuing effort to dominate and manipulate. Although he did some visiting teaching, his two preoccupations in that period were daily analytic sessions and the postwar development of cybernetics and information theory, weaving in the ideas he had been working on before the war. These were combined to provide an intellectual transition, for the first new area of research he applied himself to was the study of psychiatric communication and this in turn led into his later work on schizophrenia.

Margaret and Gregory were both intensely involved in the conferences on cybernetics and group processes sponsored by the Macy Foundation, and continued to meet there and elsewhere, working together professionally. As a child, I had little awareness of these activities. I remember our walks and the fact that he gave me a large brown horseshoe crab captured on the beach which he had had preserved and, for my first room in the downstairs of the house, a print of Picasso's painting of a boy clown. In 1949, Gregory moved to California to work with Jurgen Ruesch, work that led to their book, *Communication, the Social Matrix of Psychiatry.*

I used to fly out there to visit him in his bachelor digs. Those were early years in the emergence of patterns of visiting for the children of separated and divorced parents. In San Francisco, puzzled by the question of what to do with a little girl, he took me to restaurants and amusement parks, amazing me by the disorder in which he lived, stacks of unanswered mail and a single frying pan in which everything was cooked, with layers of lingering tastes. We spent most of our time as a twosome, although there were several years when I went to camp in Marin County where he could visit intermittently over a longer period, and he would come and travel with the camp on our pack trips. Occasionally we visited friends of his and people with whom he was working—Joe and Janie Wheelwright, Erik and Joan Erikson—and occasionally other children were found for me to play with, but my sense of those times was of being alone with him all day for several days, no one else involved, no structure or household rhythm to shape the time.

After a day or so, we would make plans and shopping lists on the back of an envelope and then pack up the trunk of his old Mercury convertible and head for the Sierras. We camped with very little equipment, a small tent in case of rain, a stove that worked with white gas, and sleeping bags on a tarp on the ground—two for him, zipped together.

The insoluble problem was my hair, which I wore in two long plaits with French braids at the sides, almost long enough to sit on. No one had taught me to care for it myself, and there was always my mother or someone else to braid it and Aunt Marie to comb it out after washing until I had it cut when I was eleven. In those days we did not have the rinses we have today, and my hair was fine and full of tangles. When I was with my mother, combing it took twenty minutes every morning and was our most intimate time of the day. She had a special stool I sat on at the foot of her bed, identifying each pull and tangle with an animal of the appropriate size and ferocity: "Ouch, that was a rhinoceros!" . . . "Sorry, darling." "Okay, Mummy, that one was just a woodchuck." My father tried and gave up, for the braids he constructed would hardly last half a day. Instead, we simply left it braided for a week at a time while we camped, letting it gradually escape into a halo of tangles around my head, full of pine needles, and then for the day of reckoning he took me to a beauty parlor to have it washed and combed.

We drove and I sang, songs I'd learned at school and songs I made up. I came home from my first trip to California with a song of many verses:

> *We'll not be together in the mountains,*
> *Nor in the forests you fear,*
> *We'll not be together in the desert,*
> *But we'll be together here.*

I varied the places but always in the same framework, and my mother wept as I dictated it to her to put in the notebook of my poetry, and I looked at her amazed, saying, "But it says we *will* be together." Gregory would sometimes sing hymn tunes, horribly off key, "Rock of Ages" or "Jerusalem." For the first few years I used to carry a supply of comic books but he had a real revulsion for the idea of my sitting there reading them, and eventually I put them aside. Wind in our faces and escape to the hills where there was no need to embody a social institution.

Sometimes we would camp in one place for several days, and then we would always pursue natural history. We would find a side road with many trails leading off from it, and make a lure with bacon or sardines, dragging it along the trails for a mile or so, hoping to interest night visitors, and then we would put out bait and sit up most of the night in the car with a thermos of hot soup, ready to switch on the head-lights or trigger the flash camera if there was any pull on the string. We pored over a book of California mammals and longed, more than anything, to see a spotted skunk, the kind that does a courting dance standing on its front legs, but we never did. We went to an area of salt marshes along the coast—I no longer know where—and waited all day in a blind with a camera, photographing the water birds, and we sat for hours in Monterey watching the sea otters.

It was early in that period that Gregory became interested in the question of the nature of play and filmed the otters in the San Francisco zoo as they recovered and then lost their playfulness. The intellectual question was the question of the metamessage, the signal that identifies a particular segment of communicative behavior as "play" or "courtship" or "threat" even though the behaviors might be very much the same. We ranged up and down the coast finding the activities that might carry the metamessage that would identify them as expressions of love and closeness. The tastes of the meals we improvised stay with me, especially lunches, a loaf of sourdough bread, Genoa salami cut with a hunting knife into jagged slices, butter grown soft from carrying in the car, and pungent cheese, spread out on a few sheets of newspaper by the roadside. Gregory believed in looking at plants but was not much interested in eating them.

Sometimes in a public campground in a state park we would strike up a conversation and someone would ask and learn that he was an anthropologist and then often enough start talking about my mother, the only anthropologist they had heard of, as often with hostility as with admiration. I remember feeling him beside me, wrestling with the question of whether and how to acknowledge the connection, and when we might escape. Once, just over the border into Nevada, we stopped at a place that looked like a country fair but was devoted to gambling. He got suckered into a game, throwing good money after bad until all the cash he had brought for the trip was lost and we had to head back for the city, while he analyzed the psychology and mathematics of the game on the way. Once when we came down from the Sierras and into the central valley the flatness made me uneasy and we drove all night to get back into the coastal range, to a place where we knew how to be. For all the informality and makeshift, we did best when we avoided "civilization" and shared a narrow place and a way of being together where the keys were "teach me something" or "let's do some natural history."

Natural history was for him paradigmatic of the way adults and children could converse, moving between an engrossing concrete reality and the abstract order of the biological world. There were other things he talked about as well. We went one year to a performance of *Oedipus Rex* and I remember his talking beforehand about the background story of Tiresias who, in the myth, had been changed into a woman because he killed two snakes copulating, and was later changed back into a man for the same action. After this, Tiresias was blinded by Hera as a punishment because, when she and Zeus argued about whether sex is better for the man or the woman, he revealed, from his comparative knowledge, that it was better for the woman, enraging Hera.

When I showed Gregory poetry I had written he was indifferent or critical, and later simply ignored those areas of my work that did not mesh with his own, assuming that eventually I would see how uninteresting they were and work on the abstractions that really mattered. Margaret had continued to write poetry occasionally after they were married although she no longer thought of herself as writing for possible publication, and she said that it was his response that finally killed the impulse for her. He drilled me on the multiplication tables and we played his version of twenty questions by the hour, analyzing the shapes of thought.

He had a genius for finding creatures. In the woods he would see the loose panel of bark on the side of a dead tree and reach in and find a bat, with a ferocious tiny face, nestled to sleep for the day. I learned to hold them, testing the small tenacious claws, and keeping them from taking off into flight from the edge of my palm. We caught snakes, pouncing to grasp them at the neck, their dry and flexible bodies coiling around hand and wrist. The feel of a snake's body is still evocative of love for me and I feel a lump in my throat and want to share the marvel with my daughter. Once a bear came raiding to our camp and I kept it away by shouting and pounding on a frying pan with a tire jack until our gear was loaded in the car and the iron skillet split in half. He would turn over old logs and find the beetles and slugs beneath them, and on the beach he would shift stones in tidal pools to reveal scuttling crabs and then carefully lay the stones back in place.

A year before he died, when I was grown up, we were together on a program at the Menninger Foundation in Topeka. There was a student there who was studying prairie ecology, and she invited us to go out one day to see a preserve of the scant remaining tall-grass prairie, the kind that stretched for hundreds of miles across the center of the country before it was plowed up for corn, grass so deep it reached up to the bellies of the horses. Seen obliquely, the tall golden grass has a shimmering surface like water, moving gently in the wind, but you can work your way down the grass stems through different levels of life, just as you can study the layers of life in a rain forest, looking under the canopy along the tree trunks, at the orchids and the bromeliads. We found a spider's web, some two feet across, where the spider had bound back the grass stems about halfway down to create a kind of cup and then woven across it, waiting, large and speckled, in the center, and we experimented very gently, seeing if we could vibrate the web without breaking it, in such a way that the spider thought he had caught his prey.

We saw a field mouse. We saw a toad. We looked at the pathways of ants, way down among the roots. And I remembered Princess Ozma of Oz whom the little girl from Kansas met traveling across the Deadly Desert on a magic green carpet that rolled out in front of her and rolled up behind her— as good a symbol of the dustbowl as has been created, a symbol of how farmers saw the prairie as hostile, waiting to be plowed up and made green, with no sense of the wonder of life within it. At another conference where we were together, in 1968, Gregory used this experience of seeing an ancient and stable ecosystem at climax as a paradigm for the aesthetic response of a whole person to a whole system.

The Kansas trip was the first time I remember seeing him walk into a new ecosystem and begin to parse it, looking at the diversity of plants and the geometry of minutely varying niches among the grass stems, but this was something he had done repeatedly over the years, beginning with his trip as a student to the Galápagos Islands where Darwin developed some of his central ideas. Later he took a student seminar, with which he traveled around the world for a year, to the game parks in Kenya. But still looking with a child's eyes after a gap of years, without growing into his natural history, my vision was concentrated on animal life and flowers, barely responding to the whole range of vegetation.

Once in a state park near the coast, we had my half sister, Nora, and my daughter, Vanni, with us, two little girls who frustrated us both because they were so much more pre-occupied with each other than with looking at the woods with us. We started talking about animism, he playing with classical ideas of the relationship between various primitive religions and the way in which children see nature, I arguing that children see much of nature as inanimate background and that one must learn to attribute life to a tree or a forest, always sharply aware of the difference between his sense of the vegetable life of the forest and my own, without a botanizing English childhood.

In the summer of 1950 Margaret came to California when I was there so that she and Gregory could tell me together that they were going to get a divorce. We went for a picnic in a state park in Marin County, grilling steaks, and then they presented the decision as a logical extension of what had been true in the past three years of separation. Margaret wrote to Marie that I lay down on the ground and cried. What I remember is asking again and again to be reas-sured that they were not angry with each other, and asking whether it was possible for people who were divorced to marry each other again. And I remember walking through the redwoods with them after a while, and hearing a radio playing, a rare sound in the forest in those days, feeling that the park my parents took me to together was not the real for-est, not the kind of forest where Daddy and I used to go.

All Margaret's divorces were Mexican divorces, to cir-cumvent the divorce laws, which in most states at that time required accusations and adversarial proceedings. Margaret had some bitterness about the manner of the decision, because Gregory had telephoned one day abruptly, demand-ing a divorce as quickly as possible in order to remarry. Per-

haps that was the only way he knew to bring things to a decision, given her rejection of his anger as extraneous. She told
me later that before they married he had said, "Well, it might
last five years, but if we decide to divorce we'll decide it in
bed." It was this promise, that the relationship would be ten-
der until the end, not any expectation of fidelity, that he had
broken. The orchestration after that, and the way in which I
was told, bears clearly the mark of her planning, and I heard
nothing of remarriage until the following winter. She had
wanted no financial settlement since her own career was far
better established than his, but wisely her lawyer suggested
that Gregory settle on me in trust a portion of his own patri-
mony so that there need be no rivalry over inheritance
between me and any children he might have later. She grieved
in secret.

I was eleven when Gregory married his second wife,
Betty Sumner. I went to California to visit them and to meet
her at Christmastime and they picked me up at the airport to
take me out to their little house in the suburbs of San Francisco. In the car, as we drove around looking at the different
firehouses scattered through the city, going from one to the
other to see the Christmas decorations, I quizzed them about
the house: How many rooms? What color were the walls?
Neither of them could tell me the color of the roof as I constructed in my mind the kind of house that children draw,
with a peaked roof, color me happy. And indeed, the house
seemed to me to be a model of the kind of domesticity that
was being held up to Americans in those days. Betty had
done secretarial work during the beginning of Gregory's collaboration with Jurgen Ruesch, but now she stayed at home
and kept house.

That day in the car, there was a new puppy, and I was

invited to name it—Osiris, the Egyptian god of the dead, about whom we were reading in school. Later he became Rusty, a mysterious name for a white dog with a few brown blotches, and the first of a series of Bateson dogs. Having a dog was one of the things you could do if you lived a settled family life in the suburbs and mowed your lawn. It seems to me that there was nothing old in that house and nothing very personal. Betty called my father Greg, which symbolized a kind of domestication, an Americanization. When my brother, John, was born in April, they completed the picture of an American family, and my sense of the Batesons at that time is held in my memory of a set of photographs I took with a camera that Gregory gave me that Christmas—dog, baby, green grass.

Betty reached out to me in the way that mothers were trying to achieve then with adolescent daughters, emphasizing a common femininity, with presents of cosmetics and cologne, and it all seemed quite wonderful, meeting needs that I had only recently become aware of. I had shifted to a new school the previous year and was acutely conscious of the differences between my family life and those of others, living in Greenwich Village and having a professional mother. There were many aspects of life in California that were omitted or quietly disapproved of at home in New York, not the least of which was a television set. Gregory's accent became gradually less English, and neighbors and colleagues were invited over for backyard meals, when new recipes were tried for salad dressings and barbecue sauces from magazines. Margaret had tried to find ways of perpetuating Gregory's Englishness, sometimes running afoul of his own rebellion against his childhood culture and sometimes creating an artificiality that seemed manipulative, but Betty was constructing around him a New World model of domesticity.

Still, Gregory made sure that on each of my trips to California there would be time in the woods together, and he gradually added more equipment. We ranged farther and the ideas discussed became more complicated. While he was working on schizophrenia, he played me tapes of sessions with patients, talking to me about metaphor and about logical levels, and in Palo Alto he took me to sit behind the one-way mirror and watch family therapy along with the psychiatric residents he was training.

In 1953 Betty gave birth to twins who lived briefly and died. In the years after that she had a series of miscarriages. The color snapshot begins to blur at that point, becoming less and less the magazine picture. There were real financial worries, because Gregory's base was at the Veterans Administration where he had been doing research on alcoholism, and this was tenuous as long as he was not an American citizen. There was a grant in 1952 with which he put together his research project on schizophrenia, then a hiatus in support followed by another grant, and I remember moments of family conflict in which grief about the dead infants and financial worry and alcohol were intermingled.

At the same time that the family drifted away from the kind of vision of domesticity that it had seemed to represent in the earliest years, Gregory began putting more of an imprint on it, asserting his own style. They moved to a new house on Colby Avenue—a real house, not newly built from plasterboard, with stairs and a fenced and private yard with live oaks and brambles as well as lawn. Shortly after that Gregory brought from England the books and furniture and pictures from the Bateson estate that had been in storage, as if deciding finally that he was ready to deal with his child-hood and youth.

I had seen some of the Bateson estate before, while it was still in the care of friends in England. After the war, my mother had brought some family jewelry to America. Much of it was mourning pieces, lockets and brooches with skeins of hair, and there was a necklace of seed pearls. There were two seals bearing a bat's wing and initials like the onyx mourning ring Gregory wore for years, and these I later passed on to John and to Gregory's stepson, Eric, and there were medals that my grandfather had received from various scientific societies. In general, the whole collection carried a message of seriousness. Gregory once commented, when we looked at an embroidery of animals his mother had done, that it was a very little bit of playfulness in a great deal of earnest effort.

Then in the spring of 1951 I had gone with my mother to England to stay with the Barlow family, old friends of the Batesons, in the village of Wendover. I remember Lady Bar-low, Aunt Nora, the granddaughter of Charles Darwin, deal-ing valiantly with the postwar lack of servants, wearing tweeds and keeping track of a large garden and a large Victo-rian family and huge house, Boswells, filled with works of art and especially with Sir Alan's collection of Chinese ceramics that later went to the British Museum, and the artwork col-lected by William Bateson.

In the dining room at Boswells, where big English break-fasts were set out in chafing dishes and the family straggled in, I remember seeing over the sideboard the Blake water-color *Satan Exulting over Eve*, which later came to Colby Avenue. Eve lies swooning on the ground, a partially eaten apple in her hand, and Satan, his wings widespread, flies a little way above her, his naked body stretched parallel to her own. Satan's other aspect, the serpent, is coiled around Eve, resting a heavy head on her breast, and flames rise around them. The painting, however finely drawn and however lovely the colors, was painful and difficult to live with. "Satan just got what he wanted," I said to my father. "Why does he look so sad?" "Because he has started the process that produced congressmen and schizophrenia and picnics and policemen on the corner, and the whole mixed bag of tricks called culture, and it's that vision that gives him the look of agony." Then he pointed out that Satan, being an angel, has no genitals, so the serpent is a part of his body, and talked about the painting as having turned up in his dreams for years.

A massive Sheraton breakfront, huge chests of embroideries, boxes and boxes of books, and artwork of other kinds collected by William Bateson also came, dissonant with the contemporary furniture and carrying a message of ambivalence into the house. Then, after a bit, the stamp of good housekeeping faded, there was more miscellany of aquaria and cameras around, and eventually after some stormy years Betty moved out and they were divorced in 1958. In the process, I saw some of the conflict I had not been allowed to see when Margaret and Gregory were separating and, because the previous history and the figure of my mother were symbolically important to both of them, heard more about that period and its complex interwoven relationships than I had ever heard at home. I also got a sense of the way my mother continued to be concerned with Gregory's affairs, often helpfully, supporting a particular grant or writing a key letter at moments when Gregory's career seemed to be floundering, but at the same time continuing a presence that could be felt as proprietary or manipulative, her name and face recurring in the media. When Gregory responded to my questions about the conflicts of that period I thought, with horror, that he sounded like the patient tapes I had heard.

In the work that Gregory did in the fifties, the concept of schizophrenia as a logical disorder created by the contradiction of logical types that Gregory and his group called the "double-bind" became linked to the concept of the schizophrenogenic family, the mother a "double-binding bitch" and the father standing aside. Gregory had darkly complicated feelings about women, starting from his own mother whom he had heartily wanted to get away from, but some of the elaboration of the schizophrenogenic mother seems to me to have been an expression of dislike of American culture and of the role of the woman in the American home of those years, the role described as Momism, a trap for her which became an enveloping trap for her children. Satan looked out of the picture frame at the Bateson house in the suburbs, at the goodwill and the care and the effort to find the right proprieties. Betty and Gregory had added on a family room, with picture windows and redwood paneling, and after Betty left, this room was gradually filled with saltwater aquaria in which Gregory kept small octopuses, wondering how their nervous systems, so different from any of the chordates, would affect the logic of their relationships. They seemed to me, as they moved gracefully through the water, like gently drifting lace handkerchiefs, even when they were engaged in territorial conflict or rivalry for mates.

Much of Gregory's courtship of his third wife, Lois Cammack, whom he married in 1960, was organized around those octopus tanks, for they used to go together to the Pacific shore to catch additional specimens and to bring back supplies of fresh seawater, brought together in common caring. Octopuses were messy but alive. Gregory once told of how he had taken one of his patients on a home visit to a rigidly tidy house and then presented the mother with a bunch of flowers, saying that he had wanted to bring her something both beautiful and untidy (he had in fact considered, at least as a metaphor, bringing a puppy). The mother parried his implied criticism by saying, "Those are not untidy flowers. As each one withers you can snip it off."[1] Again and again he made his commentary on American life by introducing that which was messy, lovely, and alive. It was an aquarium that Gregory had chosen to leave with me in Perry Street when he moved to California, almost as an analog for himself, as if the unfolding life of a mixed community of tropical fish and the care they demanded of me would keep me in touch with him across months of silence. After he and Lois moved to Hawaii, the family adopted a pair of gibbons, which swung back and forth in the greenhouse set aside for their use, deceptively affectionate when they deigned to wrap long arms around my neck, and smelling of new-mown hay.

My mother cared about the human world—all human beings, regardless of culture or race. She was simply not interested in other kinds of animals, ready to pat a kitten or admire a bird, but unready to invest real time or attention, even in ordinary kinds of pets. Gregory drew the line of his concern differently, to encompass the entire biological world, often withdrawing in distaste from human beings and concerned that what makes us so distinctively human—our consciousness—is the source of our potential for destruction.

By the mid-sixties he had adopted from a little book by Jung called *Septem Sermones ad Mortuos*[2] a frame of reference for talking about the difference between the biological and physical worlds. The biological world, Jung's *creatura*, is the world of growth, adaptation, and communication, the world in which events are caused by the perception of differences rather than by direct physical impact. A plant swivels on its stalk toward the sun because there is *more* light on one side than on another—it responds to the *difference* in light, not to the light itself. In the biological world an absence—a kind of difference between the expected and the actual, which is not itself a thing—can be a cause as well. He was uninterested in the simple billiard-ball chains of cause and effect in the physical world, and refused to believe that biological phenomena could be fully explained by reduction to a series of physical events.

Once at a conference at Lindisfarne on Long Island, one of many occasions organized by the historian Bill Thompson that brought Gregory into dialogue with other kinds of thinkers, Bill began to rhapsodize about a photograph, taken in microseconds, of the coronetlike shape formed immediately after impact by a falling drop. To Gregory that elegance of form was of no interest at all because it was the direct and automatic result of physical forces. Living creatures, messy and beautiful, were united by similarities of form and able to respond in some degree to other living beings, and it was to living creatures that Gregory responded with love and admiration. Indeed, the quality of his affection for an octopus or a bat was similar to his wry tenderness and tolerance for me or for my brother, John, or my sister, Nora.

V

Coming of Age in New York

From 1947 to 1956, I lived during the winters with my mother in Greenwich Village, first in the Franks' big brownstone house on Perry Street and later in a house my mother shared with her friend and colleague, Rhoda Metraux, on Waverly Place. Until I was ten, I went to the Downtown Community School, a turbulent progressive school that Margaret worked hard to shape, a bus ride across town. After that, when she worried that although I had learned to love learning I didn't actually seem to know very much, I transferred to an excellent girls' private school on the Upper East Side, the Brearley School. During those ten years Margaret traveled a good deal, including her first postwar field trip to Manus in 1953, but still those were years when, with all sorts of improvisations, she was making a home for me to grow up in and orchestrating my education.

The two houses had complicated similarities. In both cases we had the lower two floors, a first floor with steps up from the stoop and a basement floor below, with access to a small backyard. In both, the front room on the first floor was the original living room, dominated by long elegant windows with diaphanous white curtains at one end, while the back room was Margaret's bedroom. Wherever she lived, beds were sofas and sofas were beds, so the couch in the front room was always available for a guest and the double bed in her room was always slightly camouflaged to make the room more public. At Perry Street, one passed through her room to a tiny kitchen, with an erratic oven in which steaks had to be turned around as well as over, and beyond that to a bathroom. A flat wooden shelf was built on top of the curved refrigerator to stack the dishes, for although she cooked and closed her bed in the morning, she did no other housework and all dishes were saved up to be done by the maid, sometimes carefully stacked over a full weekend like a precarious Chinese puzzle above my head.

We ate in the near end of the living room at a drop-leaf table that was pulled away from the wall and set up for each meal, with a standing lamp brought over next to it. The table, never quite level, was the one she and her brother and sisters had done their lessons on as children, permanently misshapen by the pressure of studious elbows, and the cane-seated chairs around it were bought for less than a dollar each by my grandmother at an auction. There was always a tea trolley, so that while I set up the table in the living room with Currier and Ives mats and a miscellany of family silver—Margaret never had a matching set—she would put the entire meal on the trolley to be wheeled over next to the table. The stacked dishes, the tea trolley, the dismantled dining area were parts of her formula for New York apartment life, replicated with minor variations.

Next to the table in the front room was the bureau that held my things, for I did not dress in my own room, but came sleepily upstairs each morning for a last few drowsy minutes in her bed before I dawdled into my clothes, while Margaret, always a morning person, possessed her soul in patience. On the bureau there stood a set of reference books—a Bible, a dictionary, an *Oxford Book of English Verse*, Bulfinch's *Mythology*, a book of quotations. We seemed to turn to them constantly during meals, following up on some train of thought or completing a phrase of verse. Whether we were alone or not, I was encouraged to ask for explanations of words I didn't understand, and indeed when two English children of Bateson family friends, Philomena and Claudia Guillebaud, arrived in America as evacuees during the war, my parents fined them whenever they failed to ask the meaning of a word they were discovered not to know. Some explanations might get very elaborate indeed. Reference books would come out and one or the other would read aloud—I remember a long adventure of trying to learn to read Vachel Lindsay's "The Congo" aloud that came out of a definition of onomatopoeia. The encyclopedia was more distant and had to be fetched from downstairs, but those volumes too might accumulate on the table.

In both houses, my room was the back room downstairs. Although these basements were above ground level and not damp or cellarlike (both houses had cellars below), outsiders were critical of my mother for putting her child in the basement—one floor above her would have been acceptable, but one floor below, she complained in exasperation, was illogically disturbing. In fact, my room at Perry Street was never a great success, dark and remote from the rest of the house, necessarily barred, and made darker by my concept, when my mother let me decorate and furnish it, that I wanted it to be a forest glade, so it was painted green.

The front room of the basement at Perry Street had been intended as a study for my father and when he moved out was occupied by a wartime friend of his, Jim Myesburgh, who lived there for years and was succeeded by Philomena Guillebaud when she returned to America for graduate school, neither of them carrying particular responsibility for me most of the time but peopling the house so that at the bottom and at the top where the Franks were it was occupied. The presence of the Frank household and the various people who occupied that front room of the basement meant that as I got older I could come home alone from school to a house that was not completely empty.

On evenings when my mother was home, she would arrive from the Museum of Natural History at six or so. While she was in the kitchen wearing over her office clothes one of the wonderful full-cut aprons Aunt Marie used to make for her from dark-colored cotton prints, I would be studying on her bed in the adjoining room. Most of our supplies were ordered by phone from the grocer around the corner, but on the way home Margaret might stop off at the French bakery, where we paused to sniff in the morning, or at the fruit and vegetable shop to buy a variety of lettuces. She loved the diversity and detail of the Greenwich Village shops—the way the vegetable man would look at her appraisingly when she asked for shafts of endive and curly chicory, wondering whether to risk the impropriety of plunging the endive into the center of the chicory, or the way the fruit seller asked whether the apples and pears were to be painted or eaten. The salad was the real ritual of the meal, always the same and always different, for the rest of the meal was likely to be a pair of chops and a frozen vegetable, followed by canned peaches or apricots.

This was her salad, one that I have wandered away from since my own versions are never quite the same; no one, she said, can exactly reproduce another person's salad. She would mix the dressing in the bottom of a lopsided black wooden bowl from Bali, two parts olive oil to one of vinegar, but in the spoon she would blend at least three different kinds of vinegar—wine and tarragon and cider vinegars. She would open a little jar of pimiento and chop one half piece, add a forkful of capers, a pinch of salad herbs, and dip the tines of a fork into French mustard to stir it up with salt and pepper and a little minced garlic. And always, again, at least two or three different kinds of greens in the salad, Boston lettuce or romaine or watercress. The salad came after the meal, on separate plates, tossed when it was moved from the tea trolley to the table, with cheese if there were guests.

It seems to me that she always combined a search for efficiency and convenience with the distinctive detail that would make a set of arrangements seem personal. She incorporated each simplifying innovation of frozen foods or ready-barbecued chickens, treating them as discoveries, and yet many of her solutions were maintained over the years, and always the mosaic included fragments of tradition. Some of her arrangements came from her own mother also; now when I cook in the kitchen of my cousin Madeline or she cooks in mine, we each know how to find everything, following patterns passed into a third generation. Even today, I feel guilty about putting a milk carton on the table and I never buy iceberg lettuce for the salad or cut the lettuce with a knife.

Evenings when Margaret cooked for the two of us rarely occurred more than a couple of times in a week, but they remain central in my memory. Because we were apart a great deal, Margaret paid attention to our times together, making them into times of delight, clearly planning to make up what might be missing in quantity with quality. Sometimes children who grow up in large busy households feel, for all the hours they spent in the same house with a busy mother, that somehow they never quite met, never really paused to look and listen to each other. One year Margaret took me with her to the Vassar Summer Institute where the preschool children were busy in play groups all day but had one hour a day scheduled with their parents, walking around the beautiful campus, and she commented that many of the other mothers found that hour terribly long, for they were unused to spending time alone with their child without the background rhythms of household activity and the presence of other family members, while for her it was pure pleasure. Margaret and Gregory both spent long hours with me, hours when I felt that for the moment at least each was fully mine with no other competing concern. Margaret and I talked about everything under the sun over breakfast or dinner, and she listened, attending and responding, to whatever I said. Indeed, it seems to me that I must have done a great deal of the talking, as she drew me out on the events of the day.

It is not easy for me to summarize or describe those conversations. Instead, they are built into the interpretation of my whole range of memories from those years, kaleidoscopic and yet reassuring. We had no established way of setting up topics such as the conversations that grew out of saying to my father, "Teach me something," or "Let's do some natural history," and thus I do not retain whole chapters of narrative and explanation. It seems impossible that I never said, "Tell me about Samoa; tell me about New Guinea," yet those parts of my mother's history, before the war, seemed very remote.

We had no artifacts from Samoa or New Guinea in the house—Margaret believed it was inappropriate and a potential conflict of interest for a museum anthropologist to own artifacts—and no visitors came from those places. We did have a few Balinese carvings belonging to my father, and because of the carvings and the films and photographs from Bali, Bali was the only field trip that was vivid to me as a child, so vivid indeed that I went into a light trance watching films of Balinese dancers in trance. When the first Balinese dance group came to America, including a dancer my parents had known and photographed as a child, they came to our house. The six little girl *legong* dancers all sat in a row on one small sofa like birds on a telephone wire and after I had seen them dance I went out to buy candied violets to offer them.

Mostly our conversations arose from the present, but we also spent hours looking ahead to future plans. Because of Margaret's comings and goings, there was a great deal to be coordinated. There were major trips to look ahead to and to comment on afterward. In 1947 I went with her to Austria for the first of the Salzburg Summer Seminars and there were months of briefing for that, looking back on the war in explanations of conditions in Europe and what scarcities to expect as I planned to take chocolate and balloons for European children, and preparing to be a well-mannered child for visits in England. On the *Marine Jumper*, a troop carrier slightly converted for student transport, I dictated to her pages and pages of tone poem about the moonlit night at sea, for Margaret believed that children should be allowed to dictate letters and poems and stories long after they learn to write, until the mechanics of writing are so easy that they do not inhibit creativity.

The seminar was held at Leopoldskrone, an old and bomb-damaged palace surrounded by remains of formal gardens. At seven, I was the only child there and spent my time making friends with the students from all over Europe and with the American MPs, one of whom gave me a St. Bernard puppy, which complicated life immensely. Students took me on day trips around Austria and even into the American Zone of Germany, including a trip to a concentration camp— I don't know which one—opened up as museum. The only children I found to play with came from a nearby displaced-persons camp, where we learned to get on without a common language. My mother took me to concerts and marionette shows and to a performance of *Everyman* in the cathedral square; the voices echoing from all the surrounding towers, calling "Jedermann" to death and judgment, and my mother's explanations of the doctrine of repentance and forgiveness that lies behind the play still echo in my memory. So also do the accents of a multilingual poetry reading organized by the students, when I insisted on staying up because I wanted to hear, late at night, what Russian sounded like. Conversations about experiences like these went on for months and years.

In 1950 Margaret asked me where I would like to travel next, one morning while she was braiding my hair, and without much thought I mentioned the farthest place I could think of, Australia, "because there are so many different kinds of animals." A short time later, she was actually invited to Australia, where she had visited often on her way to and from fieldwork. She accepted on the condition that she could take me with her. That lecture tour was in 1951, when I was eleven, and we must have talked about it for a year before-hand and a year after, developing a skill of living life three-fold, in anticipation and reminiscence as well as in the events themselves, noticing for future narrative. Australia was in fact less exotic than I had hoped. I remember best the schools she arranged for me to visit, ranging from a progressive nurs-ery school to a college, and including a period in a girls' boarding school—clearly the institutions in which I could be the most intelligent observer were educational ones. But I also remember New Zealand for visits to Maori communi-ties, where the pride of the old Polynesian culture was still expressed in the arrogant postures of the old men and the overwhelming carvings.

During the years I was in elementary school, there were a series of arrangements for my care when Margaret was busy or out of town, afternoons spent with schoolmates, from whose houses I would be picked up at five or six, weekends spent at Aunt Marie's. If I was sick, the effort to have some-one available besides the maid usually meant calling on Liza, my mother's sister, who was a public-school art teacher and could come in the afternoon, bringing watercolors and a jaunty sense of wildness, eccentric clothing, tales of artists and of hours spent on picket lines, and echoes of a household that when I visited it was full of siren disorder, public quar-reling and four-letter words, and newly painted murals in the bathroom. Only when I began trying to combine mother-hood and housekeeping with professional work myself did I begin to get a sense of the complex infrastructure of my mother's life, the number of people involved in looking after me in the afternoons, getting me home, coming over to cook dinner, and of the way in which my life has been enriched by the diversity of these arrangements and the different kinds of people with whom my life was linked.

Another scene of my childhood was the Museum of Natural History, which often served as a baby-sitter as soon as I had learned to find my way back up, after hours of solitary exploration, to the fifth floor. Indeed, everyone who succeeded in visiting Margaret in the museum arrived with a sense of achievement and discovery after trekking down echoing corridors of blank-faced storage cabinets that led to one more flight of stairs up to her tower office, nested beyond shelves and shelves of American Indian pottery and other specimens. This was an odyssey that might also involve convincing some guard who didn't know me that I was allowed in these restricted areas, and often I had to ask for help to find the right elevator or corridor.

Occasionally Margaret would bring me into the museum before opening time and leave me to wander alone among the fixed tableaux of life in jungle or desert or deep in the sea, or would take me out through the darkened halls, haunted with ancient or alien lives, after closing time. Once she arranged for my birthday to bring a group of friends into the preparation rooms, to see how the great dioramas are built. She daydreamed that my wedding could be held in the Hall of Pacific Birds, against the background of a great curved diorama of flamingos flying away into the sunset, but by the time I married the museum was no longer allowing the use of any of its spaces after hours for nonofficial functions.

Through the years, the museum was invaluable as her base, but her own role was ambiguous as she remained apparently indefinitely as associate curator (she was finally made a curator in 1964), like so many women in academe who were not given appropriate status. She worked around the problems, cannily expanding her work space in the tower where male curators were not interested in competing, raising her own funds, and increasing her freedom to come and go as she liked.

For years there was talk of finally doing "her hall," a new hall of the peoples of the South Pacific, but it was not actually opened until 1971, after she had formally retired—and by the time of her death it had been closed up and packed away to allow for a change in floor plan. The image she tried to build into her hall was that of a multiplicity of islands, each elaborating different cultural themes, divided by wide stretches of blue, the reaches of sea and air and sky, crisscrossed in perilous voyage. The displays were meant to be suspended in light in their transparent cases, while other areas suggested shadowed jungle. The rationale is elegant but the hall itself was disappointing, with sections of exotic material not quite integrated into a whole.

My childhood too was a time of movement between a multiplicity of disconnected settings. My favorite form of play consisted of elaborate fantasies, often a long sequence of episodes in a sustained drama shared with one particular companion in one particular setting. One series was built around the box turtle of a boy from my class, engaged in organizing endless warfare with a tribe of white mice. Another series involved glass and china animals, spread out on a girl friend's bedroom floor around a mirror lake, benevolently ruled by a white china doe. Often there were no props, just two children in a playground acting out dramas of princes and princesses or of a horse and a loving master, but sometimes the props were elaborate.

My most important playthings at Perry Street were a collection of dress-up costumes. Gregory had bought me as a birthday present a full set of professional theatrical makeup, greasepaint and nose putty and even the makings of false beards and mustaches. One year when the fashion was for large and gaudy artificial jewelry, Margaret gave me a box of jewels, "pearl" necklaces and "diamond" chokers, beautifully displayed in a box covered with cellophane. (This was typical of her favorite kind of present, a collection of some sort, each item lovingly selected to be different but to express a theme—a dozen tree ornaments for a couple's first Christmas tree, a dozen small bottles of classic liqueurs for my husband during our thinnest years of student marriage.) I used to coax old dresses from my mother or buy them from secondhand stores on the way home from school, hoarding my bus money. The very best costumes for kings and queens, however, were academic gowns and the many doctoral hoods Margaret had accumulated with honorary degrees.

I did not have a single world of fantasy as a child but several, each shared with a different playmate and geared to his or her interests and preoccupations, not unlike the times when my mother was involved in writing several books with different co-authors. The multiple lines of fantasy were a way of drawing strands of continuity across the multiple arrangements for my care, time spent in different households where in each case a modality of play had to be invented and then suspended.

All this moving around was made possible by the fact that I was usually a well-behaved child, laughing and contented, most of my mischief ingenious but harmless, involving elaborate imagination. My mother used to say that rather than her having to ask friends if they would have me, they used to want to borrow me because their own children played quietly through the afternoon when we were involved in some long fantasy or project. Now I believe that some of that goodness was my own solution to the comings and goings. A pattern of misbehavior might have been amplified by resentment and expressed an appeal to my parents—to my mother especially—not to go away, but I went in the other direction, generally amiable and virtuous, determined to be a pleasure to come home to.

When I think of crying at my father's departures, before he went away to war, I can feel his distaste and embarrassment, palpable in the air, at having to detach a crying toddler clinging desperately to his long raincoat, and I remember my mother saying, "Daddy hates scenes especially because he's English; you have to learn not to make a fuss." I remember not yet knowing how to let him go even as I was afraid that my hanging on would keep him from coming back.

In 1946, aged five and six, Colin and I expressed our rage at one of the older children at Cloverly who had snubbed us by hiding a little packet of feces under his pillow. Nothing happened for days, perhaps weeks, and then my mother came back and pressed me for a confession which I gave in installments, not wanting to tell her the full story until it was clear that she knew it already. Then she pressed me to define a penalty and an attempt at reparation, spending my entire collection of shiny Roosevelt dimes on an exposure meter as a propitiatory gift. I remember her distress as she said, "I can't be a punishing mother; I'm away too much of the time. I can't come home to this kind of thing." It seems to me that the episode must have been saved for her to deal with, like a father coming home from work and asked to discipline the children, and the role was one she refused to accept. The discussion was reasonable and logical and seemingly endless, but it left me with a lingering envy of those whose parents were near at hand and swiftly angered and appeased. Again, the solution was to be good, for perhaps if too many homecomings were marred she might prefer not to return.

The question of discipline was more complicated if it concerned school, for, like most professional women in that era, she had to defend her choices from the critical attitudes of teachers. Once when I was eleven or so I played hooky with a friend, hiding out for the day in Aunt Marie's apartment while she was away at work. I had called Brearley in the morning to say I was sick but I had forgotten that there was a school art exhibit opening that day in which a painting of mine was hung in a place of honor. My mother came rushing back from an out-of-town trip to attend the opening, and was first embarrassed to have to plead ignorance when a teacher greeted her saying, "We were sorry to hear that Cathy was sick. Is she better now?" and then humiliated when the truancy was discovered. In the end I felt terribly guilty that my small rebellion had made her vulnerable.

It was important too to be happy, not to welcome her home with complaints. As a child, it is hard to know what problems can be solved and what cannot: When a drizzle continues day after day, complaint helps little, and those spells of ill weather that are part of an educational system or a neighborhood or inflicted by one's age mates seem equally to be unchangeable forces of nature. One day at breakfast, in my second year at the Brearley School, my mother described a parents' meeting the evening before at which another parent complained about the hazing of new children. "I told them," she said, "that you had never had any problem of that kind," and I burst into tears. It had simply never occurred to me that a year and a half of miserable isolation was something I should tell her about.

It was clear to me as it would have been to a lonely black or Jewish child in an alien school that the problem had to do with who we were—a different kind of people, a different kind of family. I lived in the wrong part of the city, came to school by subway, and shopped at Macy's. I was full of knowledge and experience that marked me as an oddity and yet I lacked the really important skills, particularly athletic skills, that would have won acceptance. Too smart. And yet Margaret's very activism, her very conviction that any problem could be solved, was a reason for not confiding in her. When I finally explained my problems at Brearley, one of her solutions was to suggest hiring a tutor for sports. I can see now in retrospect that real skills could have been taught, for all my play periods at the school before Brearley had been spent on fantasy and the playacting of private dramas, not on ball games, and I had only learned the rudiments of the rules of baseball by looking it up in a children's encyclopedia. But the idea appalled me—to know by tutoring what others knew, apparently, by birthright, was more alienating than to judge oneself incapable, more humiliating if it became known than to be the last choice in every selection of teams. After that I visited a dozen other schools, including boarding schools, and finally decided to stay on as the value systems of my classmates slowly shifted and friendships developed.

I think Margaret and I must have talked about people a great deal, as she gave history and context to the great diversity of individuals coming through that house and the range of people we visited. During those years there was a series of research projects on contemporary cultures that involved the creation of groups that partly replicated within themselves the diversity of some other society, groups that sometimes met in our living room in the evenings. I seem to know a great many life histories—how this one had become a writer by reading through every book in a small midwestern library, how that one had taught English to immigrants in night school, sequences of marriages and migrations uniquely identifying individuals and their contributions, offered in explanation for their behavior.

In fact, the models for most of the commitments I have made in my life were there in the network my mother created, so that I might have built on materials I was offered any of many life-styles: marriage with children or marriage without children, marriage open or faithful, transient or sustained, homemaking or outside career, solitude or commitment, the love of men or the love of women. It seems to me that most of what she taught me about how to behave was rooted in the specifics of individuals or households rather than in general principles. She cared greatly for detail and style, appreciating individual diversity almost as a connoisseur appreciates works of art, and bringing me into relationship with a range of people who had made profoundly different choices and represented different ways of being human, different ways of constructing family life, different ways of being a woman. Under it all, and under the possibility of constructing a variety of commitments and making new combinations, she always gave me a sense of a deep and continuing romantic love for my father.

The two households in which I spent the most time outside of 72 Perry Street represented as wide a contrast as could be found among our close friends. As an adult, reading Ruth Benedict's book *Patterns of Culture* for the first time, I read with a sense of recognition, not because I knew about Zuñi or Plains Indian culture, but because experience of different households had taught me the meaning of what Ruth called "pattern" and the kind of thematic consistency my father called ethos, the pervasiveness and congruity of style within a system that make any culture more than a list of traits and institutions.

My mother used the contrast between the household of my Aunt Marie and that of Sara and Allen Ullman, whose daughter was one of my closest friends, quite explicitly to underline some of the subtleties of difference and patterning. Worlds apart, both were deeply generous to me, but generosity had very different meanings. They also had the similarity of being two households, unlike others I spent time in, where I was absolutely forbidden to go barefoot, but the reasons, the contexts of the prohibition, were entirely different.

In the Ullmans' apartment, a huge rent-controlled apartment up many flights of stairs, just east of Sixth Avenue on Waverly Place, the floors were dirty and the style Bohemian—but it was understood that in an artist's home one should be prepared to encounter thumbtacks on the floor, so shoes, if nothing else, were required. In Aunt Marie's apartment, everything was kept as clean as New York living would allow. There were antimacassars on the chairs and a child's dirty feet were a kind of rebuke. At home and in many other places I went happily barefoot, but no set of formulas about proper behavior "when you are visiting" would have covered the variations.

Sara and Allen Ullman lived a wonderfully hospitable life, inviting friends to meals and parties and then going through periods of being "stony broke" when Sara would clerk in a shop. Martha, their daughter, another only child who was later my maid of honor, went to the Downtown Community School with me, where Allen taught painting, and Sara would pick us up after school so I could spend the afternoon at their house. Then, often, my mother would arrive with a big steak and a couple of bottles of wine and a feast would be produced and shared, rich with garlic, the meat and salad and bread flavoring each other on Mexican pottery plates, the dishes stacked randomly in the kitchen for another day while everyone sat for hours, slightly tipsy, Allen telling outrageous dirty jokes and roaring with laughter or holding forth on politics. Allen and Sara would quarrel and flirt and sparkle, both resolving to start a diet the next morning.

Both of them moved in an evocation of sensory pleasure, food and sex and color, a continuing love affair of two people who just happened to be married. Margaret delighted in pointing out the way in which Sara, even as she became fat and gray-haired, still expressed the conviction of being an attractive woman, carrying herself superbly, full-busted in décolleté black dresses displaying a deep cleavage, charming and flirtatious, the kind of woman to whom friends gave bottle after bottle of perfume that cluttered the bathroom shelves. The Ullmans quarreled publicly and Allen occasionally pursued other women, but they were the only couple of whom it consciously occurred to me in my childhood that they actually made love, in the big double bed that stood in the living room under Allen's huge bright mural of a buffet.

For several years after I changed schools, I took painting lessons with Allen, going to a big studio loft in what later became the West Village, much as I took piano lessons for a while with a composer friend, Colin McPhee, who had been in Bali when Margaret and Gregory were there and was now in continual poverty. Allen was one of the world's great teachers, working at his own canvas on the other side of the room and leaving me mostly to do what I wanted, buoyed up on his enthusiasm. When the Ullman finances became really difficult, my mother would start planning if necessary to commission a portrait bust of me by Allen, but they never faltered in the confidence that one day his work would be recognized. The main painting by Allen that hung in our house, however, was on loan, and she always said she would never be able to afford to buy it—until one year when Allen gave it to her. It hangs in my house today. Allen did almost all his painting by applying the colors with a painting knife to a white ground, working in the brilliant blues and oranges.

Aunt Marie's apartment was even more central than the Ullmans', the place where I spent most of my weekends, while my mother went off lecturing or to meetings, and where I felt fully a child of the house. There too there were things my mother wanted me to learn. Just as Margaret entrusted her finances to Marie, she also let her construct a method to teach me how to handle money, giving me my first spending money, ten cents a week which I was taught to save, and then setting up with me a system for a clothing allowance. From the time when Nanny left, Aunt Marie was in charge of my clothes, making many of them and mending them as needed, plaid skirts and little blouses cut from my father's old shirts, new sweaters knitted for each winter, sensible brown oxfords bought a size too large to grow into. She taught me to use a thimble and always to pin and baste before sewing, waxing the thread if it needed to be strong.

Under Aunt Marie's supervision, I worked up over the years from an underwear allowance to an underwear and shoe allowance, and finally an allowance covering all my

clothes except a winter coat. We would start the fall by list-
ing exact quantities of each item, with approximate prices,
and then I could spend more on one and scrimp on another,
providing exact accounts were kept in a little black notebook
and the totals came out. She gave me cooking lessons as well,
the first meal being fried liver and onions, and helped me
make and shop for Christmas presents and wrap them in
beautiful careful packages with neatly folded corners, using
wrapping paper saved and smoothed from the previous years
and decorated with pictures cut from old Christmas cards,
taking infinite pains. My memories consist of hundreds of
details of skill and care, but the mood that runs through
them all is one of gaiety. We walked in step, arm in arm,
almost skipping, on our expeditions.

Aunt Marie's care was focused on conserving. She saved
the coupons from Raleigh cigarettes until the triumphant
moment of redeeming them for a handsome card table with a
top of inlaid veneer, and then she used that card table for
thirty years, its top always protected with a fitted plastic
cover. In her house were several generations of antiques, each
with its story, above all a beautiful doll called Lucille that
had belonged to her mother, family jewelry and silver, family
china, family clocks that struck the hour at night, and beds
with horsehair mattresses, each pointed out, with its story, as
something that would someday come to me. Once in a great
while my mother would express exasperation at some particular bit of thriftiness or spendthriftiness on someone's part
but rarely. More often she would construct a framework of
toleration by pointing out to me the way in which money
was carefully separated into categories by Aunt Marie or
flamboyantly enjoyed by the Ullmans, and the symbolic uses
of food.

Each of the different households I spent time in had its set of rules, some explicit, some implicit. In each, the grown-ups were told to enforce their own standards with me, whatever they were, and to punish me in their own ways if necessary. My mother was always ready to discuss reasons but she insisted, do what you are told quickly, there always might be a matter of safety. In the Frank household as a small child I was occasionally spanked and in the Ullman household the commonest punishment was isolation from my friend Martha or the gregarious life of the family—"If you girls don't cut that out, you'll be separated." Once Aunt Marie took my allowance away for two whole months for putting my feet up on the newly painted wall beside my bed and making smudges.

My mother's own commonest sanction of behavior she found unacceptable was to say that it was "boring." Complaints, whining, inappropriate demands were met by "Stop that at once, Cathy; that's a bore," cold and peremptory. There were other behaviors that were unacceptable in her presence—I could paint my fingernails if I wanted, but then I couldn't go places with her, and the same was true of chewing gum and comic books. When my friends started to experiment in secret with smoking, she said that she couldn't stop me and basically it was my decision, but I wasn't to smoke in her presence. This made the daring experiments of my friends tawdry and unexciting.

The different households were so important that each had to be woven into our celebrations of Christmas, which took place in layers and stages with different kinds of food and different styles. On Christmas Eve, we had a dinner for a small group of close intellectual and artistic friends. My mother cooked the steaks and made the salad, Colin McPhee made hollandaise sauce for the asparagus—with tremendous trauma and tension in those years before blenders—and Allen Ullman made zabaglione. Later, the number expanded to include Rhoda Metraux and her son Daniel, and Alan Lomax became part of the group, playing carols on his guitar until some of the party went out to midnight mass.

On Christmas morning at Perry Street the joint household with the Frank family would be reconstituted when I took my stockings upstairs to open with Colin. (It seemed that the only way to hang stockings from our marble mantelpiece was to have two, on a string, balanced on each side, and after my father left I went upstairs each year to borrow a pair of Uncle Larry's stretchy gray wool socks and never gave up my right to two.) Then we had a Christmas breakfast at the Franks' big table, ten to twenty people all exchanging presents. My mother weaned me from Christmas stockings by gradually increasing the proportion of oranges and nuts until it became clear that she was no longer interested in accumulating things to fill them.

Some years there was a Christmas dinner with members of Margaret's actual family, and this might involve a train trip to her sister's in Connecticut or her parents' in Philadelphia and was the time for turkey and Christmas pudding and the more ambivalent sharing of relatives, with cousins tentative after not seeing each other for months. Always somewhere in the day, there was a call at Aunt Marie's, when she would get out the "Berlin china," a precious harlequin set of antique coffee cups that reposed in a glass display case the rest of the year. Later, when the Franks moved away from New York, we started to have Christmas breakfast at Aunt Marie's.

In another household that was important in those years

there were two different styles—indeed, two cultures—to think about, and my mother pointed out the differences, an accentuated complementary between two styles linked with a sense of femininity and masculinity. Gregory had seen such a contrast among the Iatmul in New Guinea, so great that he posed the theoretical question of why it didn't increase steadily toward breakdown, in the pattern he called schismogenesis (later assimilated to the concept of positive or regenerative feedback), but was instead held in equilibrium.

Frank Tannenbaum and Jane Belo, Uncle Frank and Aunt Jane, had a farm near Peekskill, New York, where each evening two styles would be enacted. Jane was very much a lady. She had been married to the artist George Biddle and lived in Paris, and had then come to Bali with Colin McPhee, her second husband. There they were part of a group of artists and composers and collectors of Balinese art with whom Margaret and Gregory worked closely. She was extraordinarily delicate in her perceptions, beautiful and a little withdrawn. My father spoke of having been in love with her but of always feeling that she was so fragile and so unstable that one was afraid to touch her. She wrote a book on Balinese trance, and later spent many years in mental hospitals or drugged with Thorazine, her body clumsy and the vividness of the painful mix of perception and fantasy blurred. In the years when I knew her, Jane was already married to Frank Tannenbaum, a Columbia professor who had come to the United States as a working-class immigrant and was imprisoned for his part in labor protest. He later wrote on prison reform and then on Latin American politics, warm and earthy and feisty and protective of Jane.

Before dinner, having bathed and changed, we would meet for "cocktails" in Jane's drawing room, which for me meant the platter of hors d'oeuvres that always appeared, delicate antique furniture and artwork, some of which hangs in my house today. Jane had two dachshunds, and in the corner of the room was a small cabinet with a variety of fragile little Mexican and Balinese toys that I could carefully spread out. After dinner, we would go to Frank's big brown study where I curled up on the floor with his Great Dane while the grown-ups sat in deep leather armchairs and Frank got out the smelly pipes that never appeared in the drawing room.

One sat differently in the two rooms and spoke in a different tone of voice. The possibility was there of seeing the contrast as a difference in gender, but more clearly it was a way, one of many possible ways, of elaborating the difference in gender. In New Guinea, in the thirties, newly aware of the way cultural patterns are superimposed on temperament and gender, Margaret and her husband, Reo, and Gregory had experimented with a system for describing contrasting personality types by the points of the compass. Frank could easily have been an example of what they called a "Southwesterner," warm and practical and managerial, while Jane was what they spoke of as an "Easterner" or "fey": remote and self-absorbed. Sometimes she would speak in terms of bizarre fantasies and I remember simply entering into the conversation with her, as if it were a kind of play, one more of my shared fantasy worlds.

My mother never took me to Samoa or Bali or New Guinea, but it seems to me that the complex mosaic of friendships in which we moved provided as vivid an introduction to anthropology as any field trip would have. It was important to be adaptable and accept different styles, treating them with respect; it was important to know that the customs and styles that differed so greatly were coherent, that in each case there was a pattern that could be learned. One of the problems, indeed, in the context of such tolerance of diversity, was to learn that unusual behavior may be, in fact, an expression of pathology, and may be dangerous. New York City was much safer in those days, with less craziness and less violence in the streets, but it had all the unpredictability of any great city.

Once in Tompkins Square Park, where we were taken for outdoor play from the Downtown Community School, an exhibitionist was sitting on a bench. My friends and I earnestly discussed the moral imperatives of the situation: so to behave that he would not in any way feel that we regarded him as peculiar, not to stare and tease, as we had learned not to stare or tease a cripple or a foreigner behaving oddly, not to betray fear for this might hurt his feelings. Later, in the Eighth Avenue subway station at Eighty-first Street, where I waited every day after I started going to Brearley, there were occasional exhibitionists and I was surprised to find that my mother wanted such matters reported and passed them on to the police. Once when she did so, I concealed the fact that a particular exhibitionist had happened to be black, because I had been taught to reject the facile association of black skin with violence and worried that each case reported in which race was mentioned might strengthen prejudice.

In trying to convey to me the problem of knowing whether behavior was cultural—and therefore a functioning part of some human cultural system, however odd it might seem—or individual and therefore possibly pathological, Margaret told me that when she fell in love with Gregory, she wondered how much of her love was for a particular English style and how much was individual, and I wondered the same thing about Armenian culture when I met my husband.

She also told the story of an Irish student she had one summer who spoke all the time of seeing leprechauns. Margaret felt she had to suspend judgment about whether this was simply the expression of an unfamiliar cultural pattern until one day an Irish cop came up to her, having seen them talking together, and said, "You know, lady, that dame's completely nuts."

In spite of the risks, respect for difference and the expectation of finding a pattern one can respond to and fit in with still seem to me central, one of the hardest and most important things to convey to children, even as one is teaching them to work with their own cultural patterns. When Vanni was just under three, I took her with me to stay with a traditional Iranian family for a period of time, and I worried about how she would respond to the pervasive differences: She would hear only Persian spoken all day long; she would be in a dense group of human beings (there were ten children in the household sharing three small rooms); she would be sleeping on mats on the floor, eating Persian food, using an earth toilet and water instead of toilet paper; she would have to deal with the fact that my head would be covered at all times.

So a week or so before we went, I started telling her a series of stories, fit into the framework she was demanding at that time—"Tell me a story about Stacy the cow"—as my mother had used stories and worked on children's books to teach about race or sex. One day, Stacy the cow decided that she wanted to know more about the horses that lived in the other part of the barn, so she moved in with them. And she found that they lived in stalls and were harnessed up to a plow and got oats to eat as well as hay, and spoke to each other in whinnies instead of moos. And another time, another evening before bedtime, Stacy decided to get to know the chickens better, so she moved into the chicken coop. There they picked the grains of their food up off the ground and let the rooster boss them around. And through the week, Stacy went off on her ethnographic explorations with the pigs and the sheep and the goats and all the animals of the barnyard. Whenever we hit a snag in that Iranian household, I would remind Vanni: Remember what a hard time Stacy had getting to sleep when she had to curl up on a chicken roost? Remember how hard it was for her to figure out how to be polite to the sheep when all they would say was baa-baa?

In the end, I found I still had to make sure that she had half an hour or so each day more or less alone with me, and one of the things she would do in that time was to pull the kerchief off my head as if to reassure herself. She also had a continuing difficulty eating with the large spoon she was given and I had to find her a smaller one. She played with the children of the house and learned the conventions about taking her sandals on and off, always to be barefoot on the carpeted surfaces, never ever to go barefoot in the yard or into the outhouse. The mother threw pebbles at the children when they made too much noise and the father once threatened her with a big knife if she didn't sit still. She watched me and would have recognized my concern if this had been anything other than a culturally stylized and undangerous way of admonishing children.

With Vanni, I was trying to achieve early and quickly a degree of flexibility and openness that could be allowed to develop for me over a much longer period of time, for I was seven the first time my mother took me abroad. She did no fieldwork between my birth and 1953, and when she did go back to New Guinea in 1953 she did not take me with her. In fact, the best years for taking a child into the field are when the child is youngest, able quickly to learn the ways of another culture and staying close to the mother who can use her play with the other children as a basis for rapport with their mothers, but when I was that age Margaret was occupied with the war and the various postwar "studies at a distance" of European cultures. The teen-age years are much more difficult, as adolescents strive for a certain distance from their parents and depend on seeking conformity with their peers. But although I neither grew up in New Guinea nor came of age in Samoa, Margaret taught me to perceive and value differences and accentuate them as interesting in themselves, rather than as deviations from the ideal.

She went out of her way to create opportunities for me to be in contact with different kinds of people, and often when I developed a new interest would send me to someone who could present it to me more personally. Occasionally she worried that the range was still too narrow. She had assumed that there would be opportunities for me to learn another language early, but this didn't work out until I was ten or so, and eventually she decided that contact in Austria with children who spoke no English, even though I had never actually learned another language, had taught me the difference between "This *is* a cup" and "We *call this* a 'cup,' " had gotten the idea of different languages into my head so I could learn them when I needed to. She invested a great deal of time and effort in the Downtown Community School, where I went until I was ten, to make certain that the school was truly diverse, but the memorization needed for language did not fit the progressive model.

Although my school was successful in ensuring that there

were both substantial numbers of black students and some black teachers, she worried that they might be drawing on groups of black and white children who were too socially homogeneous. I came home and asked why the black children (coming from homes aspiring to the middle class) were so much more prudish about dirty words than the white children (from politically liberal and even radical backgrounds).

Margaret also worried about how few opportunities I had in the world of the late forties to see adult blacks working together with adult whites outside of school. She felt that racial differences are sometimes sufficiently striking that one cannot assume a child will simply ignore them if not instructed in prejudice,[1] and therefore real familiarity was important and not always easy to arrange. Thus, she had a moment of sharp anxiety about how I would react when I came home one day and burst into the living room to find her sitting with three fellow members of the Hampton Institute Board of Trustees, all male, all black, very formally dressed, conducting a rather formal planning meeting. I balanced on the threshold for a moment as she greeted me, looking around, holding a jump rope in one hand, and then I observed in the voice of a gracious hostess, "*I* am the only person in this room wearing blue." Not a bad recovery, building on years of emphasis on accepting difference and valuing individuality.

VI

One White Glove and the Sound of One Hand Clapping

A friend says to me that the last time she saw Margaret and Gregory together, twenty-five years after their divorce and about to appear in a symposium, Margaret was trying to talk Gregory into putting on socks before going on the stage. Why? Why did she think it important and he did not? Against the background of anthropological knowledge and of youth in the twenties, when groups of artists and intellectuals had cut free from conventional interdictions, Margaret and Gregory selected different styles and made very different choices. On the whole, Margaret taught me to do the equivalent of wearing socks. On the whole my father didn't.

Both of them lived profoundly unconventional lives but worried deeply about the nature of order, both in social life and in nature. Margaret cared about how she was perceived, while Gregory was generally content to be seen as flouting convention. Even so, at the end of his life, Gregory was groping for a morality based in aesthetics, in which balance and symmetry would provide the basis for an ecological peace.

Each of them had to find forms of affirmation and expression against a background of knowledge of the variety of ways of being human. And each as a parent had to make decisions about what customs to establish within their homes, and how to integrate those elements of tradition they wanted to maintain and those they wanted to change.

For my mother, proper behavior was important because it allowed choices and opened doors. In addition to being able to adapt to different environments, it was important to know the forms honored at large in the society, to have the skills required for propriety, and to practice those forms as a way of honoring others. Careful table manners were not expected in all the households I visited, but Margaret wanted basic good manners to become a habit. When we had to reply to a wedding invitation, I would spend hours with thick cream-colored stationery trying to get the formal reply correctly spaced on the page. When we were planning my own wedding we consulted Emily Post on every detail and then decided whether to follow her or not. I went to ballroom-dancing lessons at one of the dancing schools in New York that prepare children to work their way up through the systems of tea dances and cotillions toward the debutante season, since she wanted me to be able to come out if I was interested.

For lunch on dancing-school days, when school got out early, I went to the house of my Aunt Edith, very much a Junior League lady who embodied the adult version of the culture the dancing school was trying to perpetuate. She was gradually finding a voice in the study of autobiographical accounts of childhood—a society matron becoming an intellectual. When the time arrived, I was in Israel working on a kibbutz and would have found the whole idea absurd, but I would have been able to manage after all those tedious after-noons in taffeta or velveteen dresses. Margaret worried that of course I could learn to be a much better dancer and enjoy dancing more if I took a different kind of lesson—but the purpose of these dancing lessons was to learn about behavior at social occasions, to learn how to put up with shy and unskilled partners, not how to dance.

Clothing was also a matter of propriety. Margaret's clothes had to meet the complex standards of public life, just the right degree of formality, feminine but not too feminine, what she wore always a compliment to the people she was with. She enjoyed pretty clothes and she especially loved hats. In my childhood there were always four or five hats in the top of her closet that I could try on, kept in separate boxes, bought from Lacette, the dressmaker and milliner to whom she went for years. As a small person she felt she should wear small hats. They were never extravagant-looking, but she liked a bit of whimsy—a rose perched up in front perhaps, a bit of veiling, and, on my favorite hat of all, a bunch of strawberries. One wore hats to church in those days and indeed to all other formal occasions. I used to have broad-brimmed hats with ribbons hanging down behind and then one year a splendid black-velvet bonnet with bits of white fur.

Summer and winter, one wore gloves. My mother showed me how, when we had only a single pair of clean white gloves between us, you could be a lady with only a single white glove, holding it in your hand and observing convention while not quite camouflaging the fact that there was only a single glove there. Until her death, she always carried a linen handkerchief but let me go with my generation in the direction of Kleenex. Her handkerchiefs were bought by the dozen because she would leave them everywhere, and then often they were not returned for the MARGARET MEAD name tags that Aunt Marie carefully sewed onto the corner turned them into souvenirs.

Margaret's adherence to traditional forms in so many ways, whether this was a matter of wearing socks or carrying white gloves, was based on her sense of what makes for a rich life. Instead of choosing some minimum set of standards to affirm, she enjoyed elaboration and playfulness even in the trivial as part of what makes any tradition human and was reluctant to see any human pattern stripped down to basics. Thus language for her was not only a system of grammatical rules that could be used for communication, it was also the stuff of poetry and nursery rhymes and word games. Nor was religion only a system of fundamental doctrines or meta-physical orientation; it was finery with which to clothe the expression of feelings, a prayer before meat, a way for city people to move through the seasons of the year with grace.

She worked for years to improve international and cross-cultural communication, so at one time she was interested in the adoption of a world language and organized a conference on the question. At that time, it was becoming increasingly clear that two or three languages would emerge as the princi-pal vehicles of cross-cultural communication, on the United Nations model. She worried that whether we relied on one or several major world languages, this would still enforce a line between those who were using their own native language and those who were trying to follow and express themselves in a second language learned in school, a continuation of colo-nialism. Instead, she argued, one of the world's minor lan-guages, not associated with any great power, should be taken as a common world second language. In this way, all the effort of translation could be channeled in a single direction and the task of language learning distributed equally, while the other languages of the world would continue to be treasured as carriers of cultural diversity instead of being swamped.

When she produced this theory, my husband and I suggested that Armenian would make a good candidate because there are reservoirs of cosmopolitan and multilingual Armenian speakers all over the world and in both Eastern and Western camps. Afterward she referred to the idea in several speeches, along with other possible candidates. This made her a great favorite of the Armenian community who did not realize how their language would be changed in such a process and what it would mean for Armenian culture if the language ceased to be a private refuge.

In any case, the key element in her thinking was the notion that a real natural language would have to be used—not Esperanto, not Interlingua, not some computer construct—for only a real human language has the redundancy necessary for human communication. Artificial languages are designed by taking a set of abstract principles, observed universals or logically necessary components, and then using these as the basis for an efficient, consistent, and unambiguous system. In general, the task has been done badly, with the logic flawed and the characteristics of particular languages or language families treated as if they were logically necessary. The mistake is in the enterprise itself, however, as if human communication could be served by a system with no puns, no ambiguities, no lullabies, as if a human pattern could be constructed from first principles. In New Guinea she had observed the use of Pidgin English, now properly called Neo-Melanesian, which serves as a vehicle of communication among many different peoples who learn it as a second language, as Swahili does in Africa. She was more sympathetic to Pidgin than to artificial languages for it does provide an effective common ground, but it carries far too many of the marks of servitude and lacks the historical resources for nuanced communication.

The same attitudes pervaded Margaret's thinking about world hunger. Nutrition was the area she worked on first during the early years of the war, and she came back to it repeatedly over the years. If cultural meanings and elaborations are stripped away and food is treated in terms of universal human nutritional needs and minimum daily requirements, so many grams of protein and fat and different vitamins, the result is alien and unappealing. The information is helpful but must be embedded in a cultural culinary idiom, and if new foodstuffs are introduced, the forms for their use should be designed by cooks.

Once when I was quite young, perhaps nine years old, I had come into the city for a day in summer which I was spending with my mother's colleague Rhoda Metraux, and she told me I could have anything for lunch that I wanted. Testing the boundaries, I asked if I could have a lunch that consisted entirely of desserts, and she agreed. Then, after thought, I put together a list of courses that was both playful and responsible, organized around cheesecake and a variety of fruits, accompanied by milk, a reasonably healthy lunch with something sticky at the end. We had a lovely time shopping for the most beautiful fruits and cutting them up and arranging them, the presentation half the pleasure.

The story became a sort of demonstration of the values of self-demand feeding, the notion that a child allowed to choose her own foods will choose a balanced diet, within an idiom where food expresses pleasure and closeness. Human beings do not eat nutrients, they eat food, food with symbolic meanings, flavors, colors, and smells, food in the form of traditional dishes that fit the days of feast and fast and speak of the relationships of husband and wife, parent and child. We are still discovering the ways in which traditional human diets satisfy nutritional requirements—how the grinding of grain with a particular stone or storage in a particular pot has added some necessary mineral, how the favored combinations of corn and beans or bread and cheese add up to health.

Margaret did not believe that it would be possible to supply the needs of nutrition or of communication with a system put together from abstract components, and I think that she made the same decision about morality. She gave me the chance to know the flavor and texture of people living different kinds of lives with verve and dedication, rather than providing a set of abstract injunctions about how life should be lived. The general principles carry no conviction without the detail, and virtue is a sterile abstraction without a day-to-day flowering in art and courtesy. On those matters on which she was committed, she would be both vehement and poetic, and she never made purely abstract arguments but took echoes of tradition and poetic language as the carrier of conviction.

Thus it was that in relation to religion Margaret did not present me with a set of doctrines in which to believe, but set out to make sure that I knew what it would feel like to believe, for the great gap between those for whom faith is a living force and those for whom it is an irrelevance is not a disagreement about fact but an incommensurate way of experiencing.

When I think about the ways in which she chose to pass on her own deep commitment to Christianity in my childhood, I am struck by the fact that she rarely talked about either doctrine or personal prayer. We shared the narratives and poetry of religion: the stories of Jesus' life, reading the nativity narrative aloud each year from the Gospel of St. Luke, a child's book of psalms that she gave me for my birthday, called *Small Rain*, and mementos from her childhood like *A Child's Book of Saints and Friendly Beasts*, full of legends, mostly Irish, of saints in friendship with animals, weaned away from wildness. She arranged for me to spend many weekends with Aunt Marie, who took me with her to the Episcopal church and taught me the Lord's Prayer at bedtime without addressing the very complex relationship of participant and observer, metaphor and conviction that Margaret preferred not to try to spell out in her own faith.

She bought me beautiful postcard reproductions of Renaissance paintings of the gospels so that I could look at them rather than being bored during sermons, and as I grew older arranged that I sing in a children's choir instead of being bored during Sunday school. I think this was based on the conviction that much of religion is distorted by a misapprehension of the absolute seriousness and truth of metaphor as metaphor. The wordiness of sermon and Sunday school would both carry a distorted message, as if only prose could be true, a message she avoided for herself by dozing through sermons or by going to early morning services, since, in the Episcopal church, the early service has neither homily nor sermon. I went to church with her on the days of high drama, Good Friday and Easter and midnight mass at Christmas.

The Christian tradition was passed on to me as a great rich mixture, a bouillabaisse of human imagination and wonder brewed from the richness of individual lives, never reduced to a meager and tasteless minimum. She had a St. Christopher's medal on her key ring and preferred those Episcopal churches where the bent knee of genuflection and the hand moving in the sign of the cross would bridge doubt and separation. She never said, this is true, but instead, this is something I care about and enjoy, and taught me to follow the intricacies prescribed by tradition. She had, for example, a book of Holy Week services that had been given to her by her godmother, and on Good Friday we would go to the highest of high-church Episcopal parishes in New York. This was St. Mary the Virgin, which Luther Cressman's seminary classmates used to call "Smoky Mary's," saying that the bishop kept a car with the motor running outside his office so he could dash down and stop them from implementing some new detail of Anglo-Catholic ritual. It seems to me that she felt no deep necessity to comply in detail with tradition, but valued its completeness so that the mystery of the Passion was expressed in an aesthetic whole. The detailed wholeness of tradition mirrored a wholeness of commitment and confidence that she held at some deep and unarticulated level where care and communion were abiding certainties.

She had a confidence that the essential would be most likely to occur when it was embedded in rich human elaboration. The grace of the elaboration, the fact that it expressed aesthetic judgment as well as playfulness, in forms shaped and shared over time, would guarantee a certain integrity. She believed that decent and caring human relationships are sustained by courtesy. Thus, in talking about sexuality and about the functions of the human body, she clearly wanted me to be both proper—respectful of external forms—and free to play, pleased to be woman and unconstrained by gender. Being a female was fun. At the same time that I was taught that there were no limits to what women of intelligence and determination could achieve, I was taught to value being a woman, to value the scope that femininity gave for play and elaboration and to look forward to motherhood. I asked her once, as a small child, passing the General Theological Seminary, whether girls could be priests, and when she said no, I comfortably said, "Well, by the time I grow up they'll have invented girls being priests too."

She told me stories about the suffragettes on both sides of the family, of how my great-grandfather's wife had insisted on walking on his arm when he was vice-chancellor of Cambridge University, that bastion of male privilege. But she did not in the process play down traditional aspects of femininity, instead teaching me to enjoy peeking into baby carriages as we walked down the street together. She gave me dolls, but never many, for how is it possible to model a mother's concern with a multitude of interchangeable and impersonal dolls, rather than with one or two? When I began to be interested in makeup, she sent me to interview her sister Priscilla, who had been a campus beauty queen. She always felt that her own handling of makeup and fashion was inadequate to the sharp eye and invidious standards of women of her generation.

In the early and mid-fifties when I was beginning to go out to dances and parties, she too was blossoming into a new degree of charm and prettiness. She bought dresses by the Italian designer Fabiani that made her look lovely and we both wore layers of ruffled taffeta petticoats. There was no either-or involved, no acknowledgment that achievement as a woman would involve sacrifices or hard choices. "You just have to learn not to care about the dust-mice under the beds," she said.

Margaret's approach to teaching about sex involved a certain balance that kept sexuality from becoming a pervasive concern. That was a period when some psychoanalytically "enlightened" parents were pressing children into an early and tense sense of sexual possibility. She wanted to avoid this. When Americans first read *Coming of Age in Samoa*, they read it as a description of a society characterized by complete permissiveness and free love, and one reviewer described Samoan behavior—with approval?—as almost promiscuous. But no one who rereads the book now, in the 1980s, should come away with that impression. Instead, you find what she describes in an unused section of draft manuscript for *Blackberry Winter* as "a certain degree of sexual permissiveness but not too much, for enjoined active sexuality may be as stressful as enjoined chastity; a willingness to let children and young people develop as slowly as they wished. . . ."[1] The adolescents in Samoa say *Laititi a'u*, "I am but young," when pressed toward some form of precocity, sexual or otherwise.[2] Against that background, the young Samoan girls who remained virgins were as much beneficiaries of the lack of pressure to make choices as those who formed liaisons.

One of the things Margaret emphasized was that the young girls were free to go out or not to amorous rendezvous. She cited this freedom as a possible explanation for the low rate of premarital pregnancy, suggesting that perhaps the young girls, free in their sense of their own bodies, knew when they were liable to become pregnant and simply did not go down to the shore. This choice might have freed them from the pressure to comply with the rhythms of male desire that for many American girls shapes their participation in adolescent social life. Even saying this, she was not able to say how the young girls would have known when they could safely go, but we know now that reproduction can largely be controlled through the "ovulation method" that depends not on calendars and thermometers but on women's willingness to note variations in the texture of their own vaginal secretions—and their option, rarer perhaps, to say no, not tonight.

Margaret was determined that I should grow up with a feeling of friendliness toward my body, and particularly that I should have no negative feelings about menstruation. She was convinced that whole populations do not suffer from dysmenorrhea, or at least that in some groups the physical sensations associated with the beginning of a menstrual period are not normally identified as discomfort. She set out with considerable success to strike a consistently positive note, so she must have been deeply disappointed when I telephoned from California, where I was staying after my father's remarriage, to say that I had had my first period. I remember being mildly perplexed when his new wife, Betty, asked me whether I felt all right, whether I had any pains or nausea, and whether I really felt comfortable about proceeding on a planned camping trip with my father—wasn't I supposed to feel well? And for the first time I heard menstruation called "the curse."

By the next month when I joined my mother for the trip that took us to Australia for six months, she was able to pick up on her agenda. She labored in Hawaii to teach me to use tampons for she had found the invention of tampons tremendously liberating, hoping to bypass entirely the worries, embarrassments, and clumsiness of pads. Then I remember a day of sunshine somewhere in the country in Australia, with that lovely feeling of lightness that often comes two or three days into a menstrual period, and I skipped through a garden, saying how happy I felt, and she talked about feeling a possibility of love and birth and growth, all as kinds of giving, and about feeling a communion with nature in shared biological process. I remember thinking, "She's trying too hard," and feeling that my particular happiness was cheerful rather than mystical.

There is always the problem, in talking to children about sex, of making knowledge available without burdening them. She puzzled me, that day, by using a lot of imagery about other zones of the body, about how it feels to defecate and the pleasures of eating and kissing, which seemed simply irrelevant. It is curious to be able to look at that moment from two vantage points, myself as I was then and myself as I am now. I am sure that I responded politely, tolerant of what she was trying to do, but almost certainly she would have felt the lack of response in me as a withdrawal. At the same time, the moment remains in my memory and there seems to be a direct connection between what she said then and the conviction I have felt since that the menstrual cycle as experienced by women—and, secondhand, by men— might be one of the things that can shape our consciousness toward a sensitivity to the rhythms of natural systems.

I think there were two important underlying ideas in the way she talked about sex and the body. One was an effort to be clear and honest, using the proper scientific terms and avoiding that recurrent problem of sex education that goes round and round the central issue and leaves children wondering what in fact happens. The other was to keep a sense of romance, a warm positive glow touched with awe for the wonders of the human body and the varieties of pleasure it can give. I know that once after some foray into sex education that had ended with her speaking of sex as something to look forward to when one was grown-up, I asked if it would be okay to tell Colin about it, "so he can look forward to it too."

She loathed the fact that the familiar set of four-letter words are used to express anger, so that she would speak of intercourse as making love rather than as fucking, not as a euphemism but as a precision. Indeed, able to be frank and explicit about sexual behavior as an anthropologist must be, and with no thrill of emancipation such as many people felt and still feel about blasphemy or four-letter words, she avoided ever using them casually. She taught me as I have tried to teach my daughter not to develop habits of language that might slip out automatically in a situation in which they would be felt as offensive. How we understand our bodies, and the ways in which human beings have tried to understand the universe, are simply too important to be littered through our conversations, ugly and out of place as beer bottles in the wilderness.

I grew up well informed but rather solemn about sex, going off with my girl friends to share information and heap scorn on the sniggering little boys who told the silly dirty jokes children tell, who were so immature about such a serious and important matter. Over the years as I grew up, it never occurred to me that sex had any continuing place in my mother's life, since she was not living with a husband, and I think the discretion she wished to preserve put everything at a distance and meant that little was actually conveyed of sex as playful. When I was a teen-ager she would make comments by which she was trying, it seemed to me, to make sure that I knew it was all right to enjoy sexuality in a variety of ways. By that time I had read enough books not to bother to pick up the bait—after all, questions of sexuality and gender were one of the main topics of the household and I browsed in prepublication copies of the Kinsey reports. Good taste was equally important, however, and that included a certain reticence.

Both my parents were inclined to see sex as a natural expression of intensity of relationship. After the 1968 conference on conscious purpose and human adaptation, when Gregory and I had first worked together on ideas as adults with excitement and intensity, as we traveled around Austria and into Switzerland and did natural history in the woods and fields and talked all day, my father looked at me rather meditatively one day and said, he supposed . . . that really the only reason we shouldn't go to bed together . . . was the danger of genetic damage if I should get pregnant? And I said, equally low-keyed, that I thought there were other reasons too. The question was such a mild one that we could continue the conversation about culs-de-sac in the evolutionary process with only a momentary tension.

Gregory stood outside the conventions of this society, but he was compliant about conforming to certain social expectations if other people seemed to care about them. Thus, he was willing, at the time of my wedding, to come to New York and play the ceremonial role of the father of the bride. He wore a rented morning coat and stood next to Margaret in the reception line, but the details did not interest him as they interested me and my mother.

In planning our wedding, my fiancé and I had picked the flowers for the bouquets and corsages with the most elaborate care, but in the last-minute fuss Margaret arrived at the church for the only opportunity of a lifetime to play the role of mother of the bride without her corsage. Then, as she waited in the front pew, a trim French maid in uniform brought her a gardenia, and pinned it on her dress. Clearly, she said, it was an angel, sent especially to make sure that all the details were right, for although we found out afterward where the gardenia came from (it was being held at the door of the church for someone else), we never found out where the maid came from. It was the same angel perhaps that my mother felt looked after her so often, for she felt she was lucky in all she did. She was fortunate in having a guardian spirit who shared her pleasure in the details of ritual and celebration, and who felt that such details were important in the orderly running of the universe. My father played his part good-humoredly and then, halfway through the reception, he and I slipped out of the crowd, to go for a walk together along the New York street, just as we had slipped away the day before with my half brother, John, to go to the Bronx Zoo.

Gregory's way of seeing was concerned with pattern, but with pattern in a very different sense from Margaret's. In his later work he emphasized his concern with "the pattern that connects"—that connects all living beings in formal similarities of growth and adaptation, the dolphin and the crab and the flower, and by which they are united in ultimate interdependence in the biosphere. To see these patterns at all you may have to ignore a tremendous amount of superficial diversity; the reality that concerned Gregory was mathematical or nearly so. It did not, finally, matter greatly to him either whether he studied the Iatmul people of New Guinea or schizophrenia or the learning of dolphins, for the same kinds of formal pattern could be discerned in different bodies of data. The proprieties of life, and the details of custom, whether or not one wore socks for public occasions, answering letters, remembering birthdays—these seemed to him increasingly irrelevant.

Gregory's interest in abstraction does not, however, when you pursue it carefully, in itself provide an intellectual basis for his unconventionality. Intellectually it was critical for him to see forest, not trees, and yet the forest is made of trees and the details of custom that make up the fabric of life are essential to continuity. Gregory moved through intellectual paths of simplification and abstraction, but in principle he believed that only a fully elaborated cultural system can be stable.

He spoke sometimes of "eternal verities"—such truths as that two and two make four, mathematical formalisms and scientific laws in their most abstract form. The two areas of theory with which he was most concerned, cybernetics and information theory, are highly abstract and formal, but they are tools for thinking about living systems. Cybernetics was for him the study of the ways in which a system, perhaps one with many parts, can sustain a complex process so that irregularities are corrected for and the system remains within certain parameters: Thus, with a thermostat, the temperature of a room is kept "constant"—within a specified range—by constant fluctuation. Similarly, it takes a constant adjustment, involving many processes, to *govern (cybernetics* is derived from the same Greek root as the word *government)* the relationship in a forest between populations of predator and prey or the levels of sugar in the blood.

All such processes involve information transfer, the communication of "news of difference." In the room whose temperature is controlled by a thermostat, some instrument must record the temperature and feed back the news of a change to trigger a corrective change in the heating mechanism. Similarly, the information of a drop in my blood sugar level must be translated in a way that sends me out for a sandwich. Gregory's way of looking at a conversation was much like his way of looking at a forest or a lake: a system sufficiently complex so that the question of what sustains it at all and by what process of self-correction participants continue in given roles is at least as mysterious as the question of how change can occur.

To illustrate the nature of cultural stability and adaptation, Gregory sometimes used to draw examples from the British tradition, particularly the university tradition, of his childhood. He used to tell a famous story about the beams at New College, Oxford: When the time came to replace them, it was found that the trees for that purpose had been planted generations before in a cycle of self-correction a century long. It is not simply an incidental of narrative style that Gregory used to describe the conversation the college officials had with the chief forester in a way that evoked the class system—the forester enters, tugging his forelock, and says, "We was wonderin' when you'd be askin'." The continuity of class relationships, expressed in details of speech and dress, is an essential part of the stability, something that was for Gregory both hateful and beloved, for even though he found stabilizing custom good in principle he was in rebellion against his own childhood culture. Gregory often quoted Blake: "He who would do good to another, must do it in Minute Particulars. General Good is the plea of the scoundrel, hypocrite, and flatterer,"[3] but his attention to the particulars of human relations was sporadic at best, and generally he was skeptical about doing good at all.

In an important sense, his own loss of "minute particulars" by his immigration to America left him in a contradictory position without a detailed self-sustaining tradition to affirm. Having repudiated the patterns with which he grew up, he never fully adopted any other set, sometimes passively accepting the patterns of life adopted by his various wives, who worked around his indifference or recalcitrance as best they could. Eventually he adopted a critique of modern Western culture, especially in its American form, that gave him a basis for regarding most of the minute particulars that were of concern to others as trivial or perhaps as destructive, for he argued that the notion of body-mind relations that has characterized Western culture since Descartes was so wrong that any culture built on it was inherently unstable and liable to move on accelerating paths of self-destruction. It was as if he saw the whole system as miswired in such a basic way that it was unusable, and so he camped out within the culture without valuing or protecting its forms.

The situation he depicted is something like the practical joke that can be played using a dual-control electric blanket. If you reverse the position of the controls, the first attempt by either person to make an adjustment will set off a cycle of worsening maladjustment—I am cold, I set the controls beside me higher, you get too hot and turn your controls down, so I get colder, and so on. The attempt to correct actually increases the error, in a process that is called in cybernetics regenerative feedback. The adoption, through a long process of philosophical speculation and filtering into the assumptions of "common sense," of the dichotomous ideas of mind and matter, thought and emotion, man and nature, was for Gregory like the miswiring of the electric blanket—once the wiring is in the wrong place, efforts at change are palliative or worse. They may even do damage, as, in Gregory's view, feeding the hungry simply leads to the procreation of more people to starve and increases both the suffering and the damage to the ecological substrate. The intellectual task he set himself involved challenging this whole set of assumptions and seeking for new (or very ancient) ideas that would function as true premises so that humankind in relation to nature becomes in fact a single self-correcting system, not one bound for destruction. This is the primary task, with a secondary task of fending off nuclear destruction in the meantime and preserving those forms of consciousness in the arts and in religion that continue to express a different mode of thought.

It was this profound skepticism that made him so disconcerting as a member of the Board of Regents at the University of California. At the time of the Bakke decision, he characterized as insignificant the question of whether one more mediocre mind in a black- or white-skinned body would be educated in a self-serving medical materialism, and urged the Regents to attend to the question of whether the university could be a place in which progress was made on fundamental questions and a system of education developed built on the abstract patterns that connect all the fields of learning. No wonder then that he was impatient with the years I spent in academic administration trying to protect and improve structures that he was inclined to think should be discarded completely. I chose as my motto when I became an academic dean a poem by e. e. cummings that Gregory loved: "Spring is like a perhaps hand . . . arranging a window . . . placing carefully there a strange thing and a known thing here . . . and without breaking anything."[4] But for him this was a poem about dealing tenderly with natural systems, not about the human structures in which we live. Western civilization, especially in its American version, was for him a window already unforgivably cluttered with jumbled and tawdry merchandise.

Where Margaret always gave the culture in which she had grown up the benefit of the doubt, Gregory did not. He looked around at the American scene, seeing increasing corruption, and used the analogy of addiction—another form of regenerative feedback—to describe many activities and institutions. Much of Margaret's popularity with ordinary people has been based on the fact that she affirmed and respected their ways of doing things, their decencies and aspirations, even when she did not herself conform. Much of Gregory's popularity in the last decade of his life was with the counterculture, those who rejected contemporary forms. But behind his rejection was an appreciation of form and a demand for mental discipline that most of his followers did not acknowledge.

It was this, I think, that underlay Gregory's attraction to Zen Buddhism and made him hospitable to its extreme attention to formal detail, even though he always stayed at the periphery. There was also a congruity between his developing view of the sacred as immanent in the mental structure of the natural world and the immanent Buddhist sense of the sacred. He first started talking in the fifties about Zen training, in which insoluble problems are posed to the student, so that the attempt to hear the sound of one hand clapping might lead to enlightenment. He compared this process to the double bind where insoluble dilemmas are posed in the context of the key relationships of childhood and may lay the groundwork for schizophrenia, but saw similarities to humor and creativity.

Finally, I think, he found in the San Francisco Zen community a community whose epistemology united their ideas and actions, providing a kind of coherence he had missed. In *Zen in the Art of Archery*,[5] there is a description of a degree of mental discipline that makes the loosing of the arrow no longer a matter of choice, separable from the archer, but an expression of his very being. Only within the context of such coherence does it make sense to worry about the concrete details of life, for Zen mindfulness emerges directly from a theory of mind: Then it really does matter which foot moves first across the threshold, which finger touches the thumb. The single flower arrangement is not a pleasant elaboration but a metaphysical dissertation as—more precisely perhaps for Gregory than for Wordsworth—was a flower or a running brook.

MARGARET AND GREGORY BEFORE THE WAR

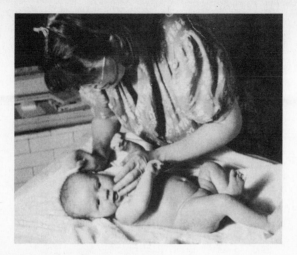

BABY PICTURES

Jane Belo

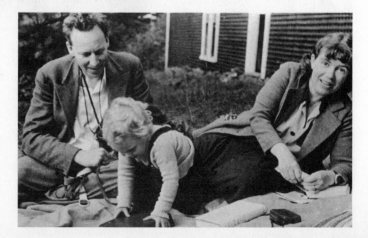

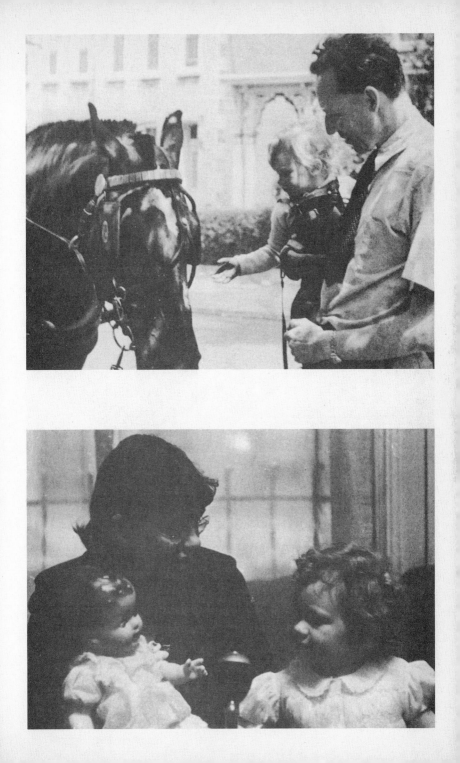

Karsten Stapelfeldt

Karsten Stapelfeldt

A FAMILY ABOUT TO DIVIDE

Above: Cathy and Colin;
Right: Aunt Mary as Ophelia

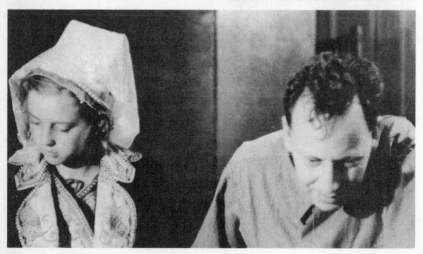

Barbara Frank as Gertrude, Gregory as Hamlet

Ken Heyman

Uncle Larry with Margaret in the fifties

Martha Wolfenstein

Lotte Jacobi

The living room at Perry Street

Paul Byers

Setting the table at Waverly Place

Joe Covello

WEDDING PICTURES

Paul Byers

Aunt Marie with Nanny

The reception line—Margaret has a broken ankle.

VII
Away from This Familiar Land

It seems important, here at the center of this account of my parents as I knew them, to trace the steps whereby I left home and to explain the unfolding pattern of my career from that time on. Over the years my parents sometimes held me or drew me back to them, and sometimes helped me launch forth with the courage or the indifference to let me fly free. Often the decisions that seemed to set a course that would take me far away have brought me back to closeness. My relationship with my father had been intermittent since he left my mother, and he had been uninvolved in day-to-day care or in decision making, so our adult relationship resembled that of childhood, a matter of visits and meetings against a variety of backgrounds. But Margaret had organized much of her life after the war around the tasks of being a parent. In the mid-fifties, however, I was no longer a child. Providing for my care as I grew older and more independent was less and less central.

She left first, going away for almost a year in 1953 when

I was thirteen to the Pacific on her first postwar field trip. By the time she returned, slim and invigorated, with a new optimism and a new set of convictions about the possibilities of elective change in human societies, I was largely out of the nest. She was away for most of 1953, communicating primarily through common letters to be shared by her family and friends, letters that arrived duplicated by her office in single-spaced purple copies, and I read them barely or not at all.

I was intensely engrossed in school that year, having finally found challenging teachers and a group of friends who shared my interests. Other factors also combined to make that year one in which I changed rapidly and matured. In November, I heard of the suicide of a boy with whom I had been corresponding, exchanging poetry and letters about poetry, and it seemed like a pivot around which my life shifted. This was not so much someone I felt in love with as someone with whom I identified, a child of social scientists who wrote and recognized my aspirations to write.

If you are away often from an infant, giving her care to others, you may miss the first word or the first unsupported step. You cannot measure your involvement by the landmarks or events that stand out, but by a judgment of the slowly evolving landscape. Similarly, you cannot know what happenstance will mark a particular day, turning a November snowfall into a pall across a year, a moment when a parent would want desperately to be at hand. Instead you must believe that the response to a moment of crisis is rooted in a thousand small events and conversations, the strength to cope or its absence determined long since, founded in the process rather than in the handling of an isolated trauma. I cannot know what my mother would have wanted to say had she been there, but I had a head full of poetry to work with and I went on writing poems and short stories about suicide, about the shattering of mirrors or falling through them, for a long time, puzzling my way through the enigma of another's finality. Although my mother and I discussed it little, I had shown her my poems and stories over the years and these must have continued to give her an index of where I was, just as she had used the poems I wrote at ten and eleven to monitor my response to her divorce. She read, as always, with a sophisticated clinical eye as well as literary interest, and would have recognized that for all the kindly adults around, I had concluded that I would have to cope on my own and that this would probably be true in the future.

After Margaret's return from the field, the Perry Street household came to an end. The Frank family moved to Massachusetts and we moved into 193 Waverly Place, reconstructing in many ways the constellations of Perry Street. We were still in the Village, but this time in a smaller, more exquisite house fronted in brick, bought by Rhoda Metraux, who lived on the upper floors with her son, Daniel, and a Haitian housekeeper, Tulia. The times were changing and in the years that followed my mother was less in New York, traveling more, more often on television, and less and less making the downstairs a separate home. In my junior year of high school I applied for early admission to Radcliffe and was rejected, but I was ready in my turn to leave home as well as school, waiting to discover when and where to go.

In the summer of 1956 I went with Margaret to Israel where she had been invited as a consultant on the assimilation of immigrants from different cultural backgrounds, particularly Oriental Jews, an opportunity to look at the contrast between real cultural disparity and the symbolic expression of unity. Young people were found for me to spend time with and I was soon traveling with them around the country, visiting kibbutzim of different types and listening to passages from the Old Testament and from modern Hebrew poetry. After ten days I tried to reach Margaret in Tel Aviv by phone and then dictated a message for her to a hotel desk clerk in the half dozen words of Hebrew that I had picked up (but could not have written), *"Shalom, Ima, ani biyrushalayim,"* "Hello, Mummy, I am in Jerusalem." When she came, the next clerk on duty, knowing who she was, looked at the message and said that it must be a mistake, but she persisted, saying she was expecting a message from her daughter. "Oh, in that case . . ." And a week later, the day before her own departure, we settled the decision for me to stay in Israel, where I learned Hebrew, entered a Hebrew high school, and received an Israeli matriculation, returning to the United States a year later to begin college.

I never lived with my mother again. I do not believe it is possible to make sense of the words I write about my parents or the ways in which I was close to them in later years without a sense of that decisive departure and separation. A quarter century later, I feel my facial muscles irresistibly shaped in a gleeful grin as I think of that telephone message and the sense of cleverness and independence it gave me, the message "Already I am at home here and beyond your ken." A quarter century later, I can pause and look at her achievement in letting me stay, the swiftness and the respect.

In 1971 I sent Margaret a Mother's Day card that said what I felt she most wanted to be told by me. It was found with her personal effects in the hospital after her death, rubbed and faded, along with the collection of family photographs that she used to carry in her purse to show people as she traveled back and forth across the United States and around the globe, as if to say, look, I do have a family, I do have roots, I am not an institution or an icon. It has a quote from William Blake, "No bird soars too high, if he soars with his own wings," and a picture of high-wheeling birds against the sky, and to it I added a few words saying that she knew this truth as few mothers do.

In *Blackberry Winter*, Margaret speaks of her fear that she might become, after her series of miscarriages, an over-protective mother. She forced herself to stand silent, feeling a painful tingle in her calves, as I set out to climb the tall pine trees behind Briarfield, after my father had boosted me up to the lowest branches. But a parent can tell whether a child is basically prudent, as I could watch Vanni, when she began to walk, almost visibly calculating each new venture, trying things that were hard for her but not too hard. I can remember Gregory telling me how to test each successive branch before putting my weight on it, and how to turn and check, as I went up the tree, the way back down.

In 1925, overriding the cautions of her professor, Franz Boas, and her father's attempts to constrain her to the ordi-nary, Margaret had gone alone to Samoa. This kind of immersion in a different cultural system and the recognition that the difference is orderly, a reflection of patterns of ele-gance, is for an anthropologist the starting point of insight. Staying on alone in Israel and throwing myself into the learn-ing of Hebrew were that for me, because of a childhood of having pattern pointed out and learning to move into con-texts of difference.

I believe that my mother was ready for us to go our sep-arate ways, that she was very much undertaking a new stage of her own life, a life that was exciting but fragmented, cen-trifugal and public. She used to speak of "postmenopausal zest," of the energies and social roles of women whose child bearing years are past and who then move more freely in society without the constraints that being a woman put on them, describing how in some societies the old women sit chewing tobacco or betel nut, trading bawdy comments with the men and rich in knowledge and tradition. So each of us was ready for disengagement. I proposed the idea of remaining alone in a new country as almost self-evident: She being who she was, I being who I was, there was no way she would not agree.

There was almost no time for arrangements. Some of the people who had been involved in her consulting trip volunteered to find a family for me to live with, to introduce me to a Hebrew teacher, and to help work out the mechanics of entering a school. She set two conditions: One was that I would write to the colleges I was interested in and make sure that staying in Israel, whether I actually succeeded in graduating from the Israeli high school or not, would not prevent my going to college the next year; the other was that if there was a war I would come home. There must be a way to climb back down. I would live in Jerusalem but friends in Tel Aviv would be "in charge" of me, and it was from their apartment that I sent a cable when the Sinai campaign broke out, saying that I wanted to stay.

By that time I had already moved out of the household where I began and had rented a room independently like the many university students in the city. These next steps had a quality of inevitability, since once I was established there it was clear that I would make my own decisions. Israel was a society of independent and autonomous youth, passionately involved with a range of ideologies, defining their own sexual and social mores in the face of the constant danger of war.

For all the links to ancient tradition, which absorbed me at least as much as the present, Israel carried the message that one could choose one's future. Margaret used Israel later, along with Manus, in her analysis of the generation gap, *Culture and Commitment*, to understand societies in which the young learn from each other because their elders are also new arrivals and cannot provide complete models. For me, a Gentile, filled with an adolescent cynicism about American culture at the end of the fifties, with only a tenuous grasp of any religious tradition, it was the model of commitment. I envied my companions and classmates who actually felt they belonged somewhere and affirmed their Jewish and Israeli identities as absolute.

The wonder was that I came back to the United States at all. My mother waited the following spring, in the aftermath of the fighting, as I hitchhiked around Israel with classmates who had finished high school and were about to enter the army, worried that I too would decide to enlist and determined to prevent me from risking citizenship and preempting the future. Indeed, I might have stayed if a firm and gentle lover had not kept pressing that a return to America and college were the next steps in my life, that the season with him and with his country was over and my place elsewhere. By that time too, the habits of cosmopolitanism and relativism had reasserted themselves: I had traveled to the Jordanian side of the border at Christmas and visited refugee camps, and wanted to understand both sides of the conflict; I had met Arabs and listened to Arabic, recognizing the cognates of Hebrew words I knew and permanently captivated by the sounds of an un-Europeanized Semitic language.

When I left Israel, I thought of myself as having completely separate interests from either of my parents, as involved in a world that was unknown and unintelligible to them, but I puzzled about their world and about my own generation. In Paris I hesitantly dipped into luxury as an old friend took my mother and me out for oysters and champagne and then my mother took me to buy a complete outfit, including lace-trimmed lingerie, to mark the contrast with the spartan life I had been living. We made a transition by talking about commitment in Israel as a social phenomenon, and later that summer I talked about Israeli youth movements at a meeting of the British Association for the Advancement of Science in Dublin. Arriving at Radcliffe in the fall, I felt estranged and let down, as if the high point in my life had passed, while other freshmen all around me were filled with excitement, and I moved to act on my one certainty, my interest in learning Arabic, which meant talking my way into a small oversubscribed class of a dozen graduate students.

My father treated all this as largely irrelevant. When I saw him briefly in October he talked almost exclusively about his own work, but he did give me to read Doughty's *Arabia Deserta* and Rebecca West's *Black Lamb and Grey Falcon*, a book about the peoples of the Balkans, two works that by their rich descriptions of Near Eastern peoples had triggered his interest in ethnography.

My mother listened to my narratives and explanations, the excitement I brought to our meeting in Paris and the sense of insight learning Hebrew had given me. Then when I left New York for Cambridge she gave me a copy of *Language*, by Edward Sapir, to read on the plane. She did not mention then that she had known Sapir, had been half in love with him and closely involved professionally, or indeed that

Sapir was an anthropologist. This was a book about language and language was what I had become passionately involved in, so that after reading Sapir I was convinced that while I would go on with my interest in the Middle East, I would also study linguistics, a field, I thought, quite separate from anthropology.

A scant month later I was introduced in a Radcliffe living room, because of my interest in Arabic, to a young engineering student from Syria, Barkev Kassarjian. My mother had called me the day before, one of a dozen friends and colleagues across the country to whom she telephoned, asking us to collect reactions to the launching of the first sputnik. All day I had been hearing analyses of the impact on the balance of power or obscure comments that suggested that the space shot was a violation of something sacred: "Man doesn't belong out there in that darkness"; "I feel that now I know the nature of sin." Barkev, who proved to be not an Arab but an Armenian, was the first person I talked to who said, simply, that he found the space shot exciting as a human achievement. The relationship prospered, filling my freshman year and banishing the sense with which I had come back from Israel that nothing would ever be exciting again and I was only going through the paces of American college life, surrounded by juvenilia.

When Barkev and I had known each other for some four or five months, we went out to dinner with my mother and then went to hear her speak at Boston's Ford Hall Forum. The topic was the future of the family. Back in Barkev's rooming house later that night, I started talking about my parents' failed marriages, my sense that with such role models I would never possibly be able to sustain a commitment to a marriage. I wept and Barkev comforted me and somehow by the end of the evening we were discussing when and whether to announce our engagement.

We were married two years later, after I learned to correspond in Armenian with my future in-laws and to cook Armenian food, and after Barkev had learned his way through the complex network of fictive aunts and uncles I grew up in. A new Armenian typeface was designed for our wedding invitation so the English and Armenian portions would balance. When we married, I had just completed my B.A. and Barkev was one year into an M.B.A. at the Harvard Business School, and both of us continued in school until we had our doctorates. Even as an undergraduate, I had made arrangements for summer language study each year so I could finish early, and this kept me in Cambridge year round.

Mine was, above all, an exogamous marriage, a marriage into a different culture rather than with the child of one of those social-science households who shared a common tradition and a common ambivalence and alienation. It continued my involvement with the Middle East and offered a share in the same kind of identity I had envied in Israel, that of the wanderer who is, at the deepest level, not estranged, carrying and sustaining a sense of home. Barkev was respectful as a future son-in-law but firm as he instructed my mother not to open the car door for herself, clear in his own style and undaunted. In response to Barkev's preference, I became no longer Cathy but Catherine.

Margaret immediately began to weave connections: Although we came from different countries, she argued that there was an essential similarity in our backgrounds. Both came from families of teachers; his aunt like her grandmother had run a school (the first nursery school in the Ottoman

Empire). Even with the Middle East, Margaret enthusiastically discovered tenuous links, for my father's uncle had been a British judge in Egypt, my mother's mother's surname, Fogg, was believed to have been passed on from a Sephardic Jewish family who settled in Foggia in Italy. Gregory came to the wedding in June of 1960, to give me away and to stand beside Margaret in the reception line.

The remainder of this story deals with the occasions when, as an adult, I dipped back into the lives of my parents, gradually adding to the role of married daughter a variety of professional collaborations, participating in conferences or writing projects. Over time, the setting of our visits changed. Betty moved out of the Colby Avenue house in 1958 and Gregory continued to live there alone, sometimes accompanied by his son, John Bateson, until he married Lois Cammack, a psychiatric social worker, in 1961, and she brought her son, Eric, almost of an age with John, into the household. The Bateson family moved to St. Thomas in the Virgin Islands in 1963, where Gregory worked with John Lilly, studying dolphin communication. From 1964 to 1971 they lived in Hawaii where Nora Bateson was born in 1969, and then in 1972, after a year of round-the-world travel, the family settled again in Ben Lomond, near Santa Cruz.

Margaret continued to live in New York, but a few years after Barkev and I were married, she and Rhoda moved out of the Waverly Place house to an apartment complex, the Beresford, on Central Park West, easy walking distance from the museum. It no longer made sense for my mother, with all of her traveling, to maintain a separate household, and although we were always welcome at the Beresford, our visits meant that bedrooms had to be shifted and temporary arrangements made, so we relied more and more on occasions when

Margaret was able to come and stay with us or to stay with my Aunt Liza who had moved from New York to Cambridge.

Barkev and I spent some ten of the next twenty years abroad, first in the Philippines and then, after our daughter's birth in 1969, in Iran, returning to the United States only for visits. I learned new languages and changed professional labels, while he became more and more a social scientist. Both in the Philippines and in Iran, he was involved in the development of graduate training in management, foreign and yet engaging deeply with the cultures and with his students while I taught or wrote or involved myself in planning and consulting.

Sometimes it has seemed to me that to maintain the continuity of family life I have had to change careers four or five times in sequence, the way my parents and some of my contemporaries have changed marriages. But even though it is difficult to master a new language or to learn to use a new kind of equipment, to read the background literature of a new field and to deal with a new cast of characters, these discontinuities are superficial. Our first hegira to the Philippines made it clear that it would be difficult to maintain Arabic, much less use it as a basis for professional activity, and the balance of professional interest in linguistics had shifted from the diversity of human patterns of communication to highly formalistic studies. At that point it occurred to me that only by redefining myself as an anthropologist would I be able to combine and sustain my interest in some coherent pattern. Linguistics is included by anthropologists as one of the branches of their field, and although I had taken few courses in other branches of anthropology, I had been exposed to it all my life in a steady current of conversation, so in Manila in 1966 I started reading and teaching anthropology. After all,

Margaret once pointed out, a Ph.D. is a certification of the capacity for study and original inquiry, not a narrow professional label. Before that I used to insist that I was not going into anthropology, but in the end anthropology became for me "the pattern that connects."

Margaret had firmly resisted the idea of guiding me in any particular direction, but I came out of childhood with the conviction that the normal human activity is writing—not, as some might expect, teaching, for teaching was an occasional or peripheral activity for Margaret and Gregory. When *And Keep Your Powder Dry* came out in 1942, dedicated to me, the first copy was put in my hands. I knew at least that this was a moment of significance to be acknowledged, related in some way to the endless sounds of typewriting, and I took the book to bed with me. When I was seven, I asked my mother when I could write a book myself. A year later, when Colin McPhee dedicated a children's book about Bali to me, my mother commented that I had been insufficiently appreciative, so I wrote a book review for him that he sent on to the *Saturday Review*, and I sniffed in scorn when they paid me only a portion of their regular rate.

There were always books in progress, usually a downstairs book and an upstairs book as Aunt Mary and Uncle Larry worked together on a series of books on child development.[1] Great sheaves of proof, still on long galley sheets in those days, would arrive at the house, to be passed on to me to draw on when the task was finished. Involved in a project to raise money for the American Cancer Society, my solution was simple: I would get all the people I knew who wrote books to donate signed copies for a special sale. One of the two big bookcases in the living room was largely devoted to "books by friends," while the other contained poetry.

Beyond the idea of writing books, I had few specific ideas as a child of what I would be or do except that it would involve a lot of studying and would perhaps be in science, and that I would someday have children of my own. In high school I found myself mainly interested in poetry and mathematics and puzzled by the dissonance that others seemed to feel in these two interests, both of which were sidetracked by the experience of living in Israel but have cropped up in other forms.

Margaret succumbed once when I was a child to the question, "What do *you* think I'll be when I grow up?" and said, "Oh, you might be something like an embryologist or a crystallographer," a strangely specific response, stimulated, no doubt, by something she had been reading, and a way of saying, "I think you will be interested in working on precise, complex patterns." It disconcerted me utterly because I had never thought of crystals or embryos as something to think about, so I teased her for years about her prediction.

In fact it was curiously accurate, accurate in its emphasis on an intellectual style, with a rather accidental sense of subject matter shaped by the opportunities of a migratory career. Setting a range of different interests within the framework of anthropology gave me the context for thinking about culturally patterned sequences of behavior—poems, for instance, or conversations, religious rituals or the series of messages whereby trust is negotiated between individuals.

My first work in linguistics, my doctoral dissertation, was a study of pre-Islamic Arabic odes.[2] These are long poems that originally attracted me by their vivid imagery of horses and camels, of desert hunts and dalliances, and, in my continuing romance with Arabic, by their sonorous language. The topic that I focused upon was the way in which the poet structured the flow of linguistic forms to create continuity and discontinuity in parallel with shifts in the thematic content of the poems.

This study, not of the forms in the abstract, but of their interrelationship in performance, seems as remote as possible from my parents' work, and yet in retrospect I can see the connections. Without my fully knowing it, it was inspired by work that Ray Birdwhistell was doing on the recording and analysis of behavior in family therapy sessions, working in close touch with both Margaret and Gregory. When I was in college and in the early years of my marriage we used to spend Thanksgiving in Ray's household. His talk of what he was doing, his sense of wonder at the richness of behavior that required such meticulous recording, was amplified by his own gestures and mobile features. His data were captured on film and tape, including gesture and tone of voice as modalities of communication. My pre-Islamic odes had been preserved from oral tradition and later written down, but they had that special richness of patterning in the verbal material that we associate with poetry. Intermediate between the two projects was a project I never carried out—to study films of Arabic political speeches, to try to understand the power that goes with the Arab sense of eloquence. Because of this imagined project, I can see a relationship between what I did—the study of seemingly immutable "classic" works—and the study of transient filmed interactions.

In the Philippines from 1966 to 1968, I did some linguistic work and some fieldwork, but mainly I taught and used the opportunity to develop a teaching base in anthropology. Our first child, Martin Krikor, was born there prematurely and died the same day, his birth very close in time to the birth of my half sister, Nora. We came back from the Philippines,

stopping in Europe on the way, where Gregory had organized his Conference on Conscious Purpose and Human
Adaptation, and I spent the following year on a grant, writing a book about that conference and reading about language
learning in children. That set of interests took me into
descriptions of conversationlike interactions and games
between mother and child and the evolving ritualization of a
charismatic prayer group. In the fall of 1969 our daughter
Sevanne Margaret—Vanni—was born.

Gregory's students used to complain to him that they did
not know what his courses were *about*. There were dolphins
in them, but they were not about dolphins; there were New
Guinea rituals and schizophrenics and alcoholics and Balinese child rearing, woven together with quotations from
Blake or Jung or Samuel Butler and the challenge to students
to look at a crab or a shell and say how they recognized it as
produced by organic growth, to look at a sacrament and say
what on earth was going on. Yet it is curious that Gregory
tended to look at my own intellectual choices with the same
sort of misplaced concreteness, disparaging my work on
Hebrew or Arabic poetry or the design of universities as if
these interests meant abandoning the interest in pattern and
becoming bogged down in specifics. The choice that kept
recurring for me was not so much between the abstract and
the concrete as of the best way to express the infusion of
moral concern that came directly from Margaret: to work on
projects that would be helpful, allow better education or better cross-cultural understanding.

In 1972, a move to Iran once again separated me from
my ongoing work, this time in a project on mother-child
communication that involved film and tape and acoustical
equipment at MIT's Research Laboratory of Electronics.

Thinking about what to do next, I read the work on Iran that had emphasized mistrust and cynicism, and my impulse was to say, yes, but when *do* Iranians trust each other—surely you cannot run a complex society totally without trust—and what is it that people admire and want to become? How can cross-cultural communication be enhanced by understanding the positive vision behind the strangeness? The same kind of impulse led me later, both in Iran and in the United States, into the administration and planning, which allows the testing of a way of thinking against a variety of situations and at the same time offers a challenge to create or sustain an ordered environment in which human beings can flourish. Being an academic dean is rather like housekeeping, however, and can bog down in a morass of maintenance in which the underlying patterns become invisible.

Over the years, conversations with my mother helped me to recognize these continuities and allowed me to stand back and think about what I was doing and why. Even so, as I look back over these sequences, it seems to me to have been critical that my path separated decisively from hers at an early stage, so that our conversations took place in a context of distance. The thing she was most afraid of as a parent was that her capacity to think herself into the lives of others, imagining possible futures, would lead her to guide me in a way I would later repudiate. In any quarrel between us, the thing that would have hurt her most, because it had been said to her so many times, would have been to accuse her of dominating me or interfering with my life. This was a point of vulnerability so deep that we conspired to protect the relationship, she by refraining from advice and indeed by trying to restrain her imagination about me, I by carefully monitoring the kinds of indecision or uncertainty I shared with her. I learned to show my imagined futures to her only when they already had a degree of vividness that I could continue to acknowledge as my own, unable to say, this is your plan I have been living, your fantasy projected onto me. She had the capacity to live many lives, participating richly, reaching out in complex empathy, grasping hold of possibilities that had so far eluded the imagination of others, and so she had to monitor the dreams she dreamed for others.

In *Blackberry Winter*, Margaret again and again uses metaphors that suggest her sense of herself as directing a play, as a producer, assembling and placing a particular constellation of people. As a girl she imagined herself as the minister's wife in a country parish, organizing everyone's life. It seems clear that it was under Margaret's influence that her first husband, Luther, became an Episcopalian, leaving the Lutheran church in which he was brought up. Later he decided to enter the seminary, and it seems possible, especially since he later became convinced that he had no real vocation and left the ministry, that he had, to some degree, allowed himself to be held in a role she had daydreamed about for years, a role whose potentiality she discerned in his seriousness and kindness. Similarly, Reo was en route to Cambridge University to continue in psychology when Margaret came out of her first fieldwork, where isolation was a major problem, and the imagination of collegiality got woven into their intense shipboard conversations so that in the course of the following year Reo redefined himself as an anthropologist.

It was Gregory, more than anyone else, who lashed back at her for trying to manage his life. She once described to me sitting in a hot room with Gregory, seeing him sweating and knowing he was simply putting up with it without thinking of removing his jacket, and knowing that if she suggested he remove it, he would be furious at her interference, furious because she had recognized his situation before he had. She would see a problem and her imagination would leap to a solution; she would see a potentiality and imagine it realized, always in danger of preempting the pleasure of discovery or of bearing the blame for failure. One of the few contexts in which she allowed herself an unbridled expression of this impulse was when she was on an airplane in bad turbulence or with some apparent misfunction. Then she would indulge herself and sit and plan out in luxurious detail all the wonderful things her beneficiaries could do with her insurance after an accidental death.

The issue of daydreaming for others was one we were both aware of and both cautious about, gradually relaxing as time passed. At first it seemed that the best protection was a degree of separateness whereby she simply did not know enough about the worlds in which I moved to begin to design the future. When I first decided to continue in Middle Eastern studies, I warned her off the Middle East, saying that henceforth this was my turf, two anthropologists each with her own tribe, and I carefully did not encourage her offers to stop off in Beirut and meet my in-laws. This meant that when we talked about my other family or my work we were talking within the framework I provided, my sense of emergent possibility already present in what I said. She had close friends in Cambridge, Massachusetts, but Harvard University was one of the few institutions she had a continuing grudge against for discourtesy and exploitation of women, and so in the nine years I was a student and young faculty member in Cambridge she had almost no professional interactions at

Harvard except as very specific favors to friends and with minimal publicity. When she came to visit Barkev and me in Manila or Tehran, we would brief her and then admire the way she wove our briefings into her public performances, clarifying our thinking in the effort to satisfy her own appetite for insight, and then she would be gone. It left us strengthened in our sense of freedom and competence in the places we had found.

Margaret's versatility and mobility meant that when we were overseas she visited us at least once a year, accepting invitations she might have otherwise rejected or postponed because they brought her to our side of the globe. She came and went, busy and prolific, always wanting a share in the flavor of our life, to hear about whatever was on our minds at a given moment and fit it into the complex mosaic she was constantly elaborating. When she came we sometimes made her coming the excuse for a big party following a public lecture. She wanted to know what was happening about family planning, how the role of women in the Philippines or Iran was changing over time, the difference in ethos between institutions we taught at, to taste, to hear, to speculate.

Once in my childhood, Ray Birdwhistell said to me, "Your mother has such a masculine mind and your father such a feminine mind." I bridled, for this was in the fifties when a comment like that seemed to be a disparagement of Gregory. Margaret in those days was already a celebrity to the general public, while almost no one outside a small circle of anthropologists and psychotherapists knew who Gregory was. People who asked me, "How does it feel to have such a famous mother?," would go on to say, "And what does your father do?" Ray then continued, "Margaret is always shooting thousands of ideas out in all directions, like sperm, while

Gregory, when he has an idea, he sits on it and develops it like a big ovum."

My adult relationship with Gregory was in direct continuity with my childhood relationship—visits to him spaced out over time, focused in dialogue, sometimes embedded in the larger conversation of a conference in which we looked at or thought about some natural phenomenon or his own evolving ideas. The number of ideas was not large—he pursued a small set of highly abstract themes all his life, although the examples and parables used for teaching them changed more rapidly. It was possible to be away for a year or two and come back, stepping again into much the same deep and slowly moving river. Introverted and involuted, his thought is well described as a sort of incubation, and he was tolerant of those who came and went if they were interested in engaging with his ideas. Unlike Margaret, he used to try to draw me away from other kinds of involvement, and seemed almost to forget my existence when I was not with him. Suspecting most of what I did of being a waste of time, he used occasionally to propose the possibility that Barkev and I might come and live nearby, so that he and I could work together on the development of abstract understanding. He would shake his head impatiently, like a horse tossing off flies, when I talked of other kinds of work and other places.

Perhaps it is because I am writing in the aftermath of my parents' deaths that today I see my relationships with them as punctuated by departures. Margaret affirmed these departures, her willingness to be left as well as the importance of being able to leave, walking away past the customs desk or into a transit lounge without looking back and letting me in my turn go into worlds unintelligible to her. In 1947 she wrote a poem for me,[3] about the freedom to depart into an unknown future, but in her file of drafts I find that she thought briefly that perhaps it was really an expression of her feelings about Gregory, who was then in the process of leaving her, and experimented briefly with recasting it to refer to him.

> *You must be free to take a path*
> *Whose end I feel no need to know,*
> *No irking fever to be sure*
> *You went where I would have you go.*

Reading it, I can see that even as Margaret affirms the importance of giving freedom, she asserts her own freedom as well. The syntax of the final stanza echoes in my mind with an ambiguity as to who is leaving, for it reads to me as if perhaps it is Margaret herself who is departing, leaving the future in my hands.

> *So you can go without regret*
> *Away from this familiar land,*
> *Leaving your kiss upon my hair,*
> *And all the future in your hands.*

VIII
Sharing a Life

In 1955, my mother drafted a letter addressed "To those I love," to be sent out to a list of friends in case she should die suddenly. Many things in her life changed between that time and the time of her death in 1978, and yet the letter was never worked on or revised in the intervening years, but instead was faithfully preserved by Aunt Marie and handed on to me. Portions of the letter still speak to me now, as I think of her later life.

> I have become increasingly conscious of the extent to which my life is becoming segmented, each piece shared with a separate person, even where within the time and space of that segment, I feel that I am being myself, and my whole self in that particular relationship. Part of this is due to age; since Ruth Benedict died there is no one alive who has read everything that I have written; different parts of my work are shared with

different people with special interest. No one, neither a devoted student, nor a close collaborator has time for all of it, now. Since the break up of my marriage, far less of my life has been shared with one person, and a multitude of special relationships, collaborations, slight gaieties and partial intensities have taken the place of a marriage which once occupied so much of my time and attention. Distance now separates me from people who once were able to keep most of the threads in their hands. . . .

The letter ends with an affirmation:

I prefer a life in which each important feeling and thought can be shared with someone whom one loves, friend or spouse—several friends, teachers, and pupils. It has not been by my choice of concealment that anyone of you have been left in ignorance of some part of my life which would seem, I know of great importance. Nor has it been from lack of trust—in any person—on my part, but only from the exigencies of the mid-twentieth century when each one of us—at least those of us who are my age—seems fated for a life which is no longer sharable.

The preference for wholeness, and grief as that wholeness is imperiled, is a theme that runs right through the lives of both my parents. The note of sorrow Margaret expressed in this draft letter continued and deepened through the remainder of her life as a background to the achievements and satisfactions of the next twenty-five years. Wholeness was indispensable to her work, for all her insights into the functioning of human societies were grounded in the intimate knowledge of small integrated communities within which she could know every person, and know them as defined by a set of interlocking relationships. In her private life, however, there was the increasing fragmentation that this letter described, even as she moved in ever-expanding circles, becoming more and more an international figure.

Margaret spoke sometimes about the "anthropological sample"—not a group of respondents selected so that they might be as representative as possible, random with regard to all distinctive characteristics, but a group of identified persons who could be seen as representative only if their place in the network of relationships were fully known. In the context of the system, the Queen of England or the prostitute on the corner could each be seen as representative of specified positions within the social fabric, and their idiosyncrasies seen as fitting in. No one of us has such a complete picture of my mother, a map of the network of relationships that was her mode of being, and thus we are unable fully to define who she was. This letter conveys her grief at becoming gradually unknown even as she became most famous. She went on to say, "I have no sense of having become inexplicable," and yet she spoke of a life which had become "no longer sharable."

Such was Margaret's vividness that even people with whom she had the most fragmentary of contacts felt they had encountered a whole personality. Their numbers increased from year to year, even as she worked to maintain and strengthen tenuous ties. I go through her Christmas card list, for which she addressed the envelopes by hand for years, usually sending photographs of herself or of me and my family, and I find scores of names I cannot identify. As time passes and I talk to others and think of what I know of her, I learn of unimagined dimensions in even the relationships I knew were most important.

A parent dies and one gropes for a certain knowledge of the person who is gone. More and more, it has seemed to me that the idea of an individual, the idea that there is someone to be known, separate from the relationships, is simply an error. As a relationship is broken or a new one developed, there is a new person. So we create each other, bring each other into being by being part of the matrix in which the other exists. We grope for a sense of a whole person who has departed in order to believe that as whole persons we remain and continue, but torn out of the continuing gestation of our meetings one with another, whoever seems to remain is thrust into a new life. Then, too, my mother created and sustained images of persons and relationships as she wanted them to be, and did so with such vividness that others saw the world through her eyes, defining themselves through her imaginations which only gradually fade.

Margaret worked hard and incessantly to sustain relationships, caring most about those in which different kinds of intimacy supported and enriched each other, the sharing of a fine meal, the wrestling of intense intellectual collaboration, the delights of lovemaking. Her letter takes the death of Ruth Benedict in 1948 and the dissolution of her marriage to my father, gradually becoming irreversible in the same period, as the end of a kind of completeness. Ruth and Gregory were the two people she loved most fully and abidingly, exploring all the possibilities of personal and intellectual closeness. The intimacy to which Margaret and Ruth progressed after Margaret's completion of her degree became the model for one axis of her life while the other was defined in relation to the men she loved or married. After Margaret's death, I asked my father how he had felt about the idea of Margaret and Ruth as lovers, a relationship that had begun before Margaret and Gregory met, and continued into the years of their marriage. He spoke of Ruth as his senior, someone for whom he had great respect and always a sense of distance, and of her remote beauty. What came through quite clearly was a sense of the incongruity of any kind of jealousy or competition.

Margaret's final tribute to Ruth Benedict was to write her biography. Through much of my childhood Ruth's personal papers, now in the library of Vassar College from which she graduated, were spread out in the living room of my Aunt Marie who systematically worked her way through them and catalogued them for my mother to use. Ruth was a woman who slowly and painfully, over many years of frequent depression and perplexity, found a path away from the traditional set of social expectations for women to her own distinctive identity and creativity. In that process she withdrew from the frustrations of a childless marriage and created out of her sensibility as a poet a new style of anthropological work. As I know more about her, I find myself wishing her well and celebrating that late fulfillment both in her profession and in friendships of different kinds with younger women, as a teacher, companion, or lover, as well as the new kinds of relationship with both men and women made possible by a clear professional identity. I get out and sort through the very tenuous memories I have of her, enjoying the fact that she indicated that her table silver should be kept for me, thinking ahead of me as an adult woman working out my own combinations.

Margaret had no such history of alienation and unmet

need. She continued throughout her life to affirm the possibility of many kinds of love, with both men and women, rejecting neither, and she went through her life not with a sense of impoverishment but with a zestful sense of asking for more, for experience enriched and intensified, sometimes exhausting to those she spent time with, impatient of all possessiveness and jealousy. Through the major part of her adult life, she sustained an intimate relationship with a man and another with a woman. This double pattern must have been satisfying and sustaining, but at the same time it created a kind of isolation, an isolation of secrecy.

The letter she wrote in 1955 has seemed to me to be an expression of concern that, due to some accident, details of her life might be revealed under circumstances of scandal or notoriety, circumstances under which she was not there to provide explanations and reassurances to those she cared about. Increasingly, however, it seems to me that the list of those whom she would have wanted to address would have broadened over time, and that real reassurance lies in a restoration of wholeness and honesty. Then too, the possibility of revelation is not something to be considered only in the case of some sudden accident but a near certainty as researchers work over her life and more and more documents become available. At the same time, the truth becomes more discernible and more tolerable as public understandings change, and we are no longer in the fifties but in the eighties.

Indirectly, she expressed these concerns. In 1957, she was posing to her friends the question of how she might disappear involuntarily and without trace. Even as we conjured up fantasies of amnesia and political intrigue, she produced a plot in which she had gone to meet an old friend whom no one knew, in a small rundown hotel that burned down. She was meeting the man who was her lover in those days in a borrowed apartment, so one can imagine that one day a fire broke out in a neighboring building, or perhaps it was only that a fire engine passed, with alarming clangor, setting off sequences of narrative in her mind. She would have been concerned about the effect on his family, but even more about people who were sure of their knowledge of the tenor of her life and of their place in it, suddenly shocked to discover that she had kept secrets from them. She clearly believed that the keeping of these secrets was correct and responsible behavior, a precondition to her availability to do work that she felt was important and to be heard in saying things that needed to be said. Today the lover of those years, like her, is dead, and the segmentation of her life, which she thought of then in the context of a finite set of relationships, seems to me a metaphor for the difficulty of sharing any life or for sustaining this common planet except through sharing.

I was fifteen when the letter was written, one of those to whom it would have been sent. I knew little until after her death of the pattern of relationships to male and female lovers that she had developed, so that trying to look back on who she was as a person and as my mother has been complicated by the need to revise my picture of her in important ways and by the need to deal with the fact of concealment. I have been at times angered at the sense of being deliberately deceived and at having been without doubt a collaborator in my own deception, limiting my perceptions to the images she was willing to have me see. I have sometimes felt myself doubly bereaved as well, having radically to reconsider my convictions about who she was and therefore, in relationship to her, about who I was and am, surprised at last by the sense of continuing recognition.

What she carefully concealed, I have now decided to write and publish. It has seemed to me finally that if we are to winnow out what is valuable and freeing in her work, it is necessary to know who she was with whatever honesty we can achieve. Children do not, I believe, belong in their parents' bedrooms, nor does the public belong in the bedrooms of those it has turned into public figures for their wit or their beauty or their wisdom. But Margaret Mead has walked in a thousand bedrooms, has been a touchstone for parents trying to understand the sexuality and sexual orientation of their children, has both helped and hindered women trying to understand themselves and their potential. Those who have attended to her words have, I believe, the right to know something of her experience, even as they realize that no one can fully represent in the single life they lead the full human potential of their vision.

We also need to understand these most intimate relationships because of what she taught over the years about friendship, for she used to say that those men and women who were most vehement in rejecting any kind of tenderness with their own sex would be incapable of friendship. She brought me up in the fifties, before women had rediscovered the friendship and support they owe each other, never, ever to break a date with a woman in order to accept one with a man. At the same time, she also talked about the need for a woman to have real friendships with men, men who are not lovers, and the pleasure of friendship cultivated in the soil of passion.

She left other material for understanding her double pattern of intimacy in her discussion of the book *Portrait of a Marriage*, by Nigel Nicolson, published in *Redbook* in 1974[1] with the title "Bisexuality: What's It All About?" In it she described the exploration of sexual relationships with both men and women as an expression of creativity, of breadth of potential and innovation, "among groups in which the cultivation of individuality has been a central value"—people for whom "the differences between men and women and the differences between individuals in temperament and gift were the basis of love affairs of great intensity and meaning . . . as part of a world in which politics, art, music, theater and intellectual concerns were complexly interwoven."

In this article, she discussed the gradual reduction in recent years of savage taboos and laws against homosexuality, but she also spoke of a tendency to limit our understanding of sexual variation so that even when we are being most accepting we tend to think of a mirror-image group of homosexuals in counterpoint to heterosexuals, equally specific in their preferences and able to find solidarity in a group or in a liberation movement sharing the same commitment. Bisexuality is most often acknowledged in the recognition that preferences may change serially, with one period of life devoted to explorations of relationships exclusively with one sex and another with the other, for we live in a society which mandates choice, at least for the moment. Even as we begin to permit differences in sexual preference, we ask individuals to align themselves one way or the other. It was this process of labeling that she rejected most of all.

In Margaret's youth it was not impossible for a woman to choose an active career, but frequently this was a choice that precluded domesticity. Margaret in general refused to express her affirmations as rejections, just as she refused to repudiate any of her marriages as valueless, and criticized newly divorced friends for the retrospective falsification that claimed, "We were never happy; it was a mistake from the beginning." She remembered each of her husbands with affection and Gregory with abiding love in spite of his rejection of her, and she worked toward friendships with each and with their wives, and also valued the sense of unrealized possibilities. There seems to me to be a connection between the openness that meant that she did not have to affirm the present good as the only good or the present choice as the only possible choice, and the openness to different ways of life and thought that was essential to her anthropological work.

Today what seems furthest from our habitual ideas is her belief in the possibility of sustaining more than one deeply loving intimacy at a time. There is a conviction in our culture that real love is exclusive, preferably for a lifetime. So deep is this belief that when we wish to defend homosexual preferences we frequently point to the possibility of sustained pairing. She said to me as a child, "You may someday find yourself feeling that you are in love with two people and think that that is impossible. If two people are really different you can indeed be in love with both." She argued with me about the validity of the idea of a "best friend" or a "favorite color," saying that each person has many aspects. In a particular mood or for a particular activity you would seek out one or another companion, just as there would be times when you would prefer blue to yellow and others when you would prefer yellow to blue. She spoke of the pleasures of similarity and the pleasures of contrast, of relationships with women as providing the comfort of the known and relationships with men as always containing the element of arbitrary and mysterious difference.

As I reread her 1955 letter in the context of an increasing knowledge of her various commitments, I keep coming back to the word *share*. One of the reasons that Margaret was so prolific is that whatever came to her was enjoyed through the acting of sharing. This was as true of an idea as of some more material good. Although she spent a great deal of time alone in her life, most of that solitary time was spent in intense and disciplined work, reading, or dealing with a stack of mail, or getting up very early in the morning to "write five thousand words before breakfast." Solitary relaxation was not something she sought and she had a virtual taboo on solitary meals, making every effort to eat with a companion, almost all pleasure defined as social. Over the years, she shared the pleasure of eating many gourmet meals and drinking excellent wines with Geoffrey Gorer, as well as visiting works of art and walking with him through beautiful gardens and galleries, but he commented to me that she hardly noticed the taste of food when the enjoyment was not part of a relationship. Rather than companionship being a drain of energy, she was always nourished by it. Even more important for her was the fertility of an interaction as a way of generating new ideas and insights. She thought into speech or onto paper, imagining when she was alone the group or the person with whom an idea would be shared.

As the years passed, Margaret had an accumulation of confidential knowledge and an increasing body of experience that could not be brought into a new relationship, and thus the self that she presented was increasingly selective. Just as no one except perhaps some future scholarly biographer could be expected to find his or her way through her mass of publications, so her task in establishing a relationship became one of finding quickly a point of contact or common interest. Increasingly that point was located in the other person's experience or work. For myself, it seems to me that like many children I set limits on my curiosity about my mother, freeing myself for the tasks of my own life. I have never tried to read all her writings and indeed have only read about half of her publications after I was an adult and an anthropological colleague—it was too important for me to acquire my own point of view. But she read everything I published, following the various literary magazines I worked on in school, sometimes reading several drafts, wading through a dissertation and through technical linguistic discussions as my interests slowly moved closer to hers.

At a certain point, when I was in my mid-twenties and she in her sixties, she apparently decided that it was important that I should know that she had some continuing love life, lest I fall into the assumption that constrains many Americans in all their intimacy after early middle age, that sexual activity does not mix with gray hair. She must have considered, as mothers do in trying to discuss such matters with daughters, how much to tell, what would be comfortable and what useful, and told me briefly of the man who was her lover for many years—skeletally, with no details, no names or places, the identity carefully masked by the falsehood that he was a retired military man with no relationship to her except for the liaison and no interests in common except the delights of a particular occasional intimacy. The careful emphasis on a relationship with no content other than sex struck me as improbable and faintly vulgar—I could not imagine any important relationship in her life that was not awash in intense conversation—and there the confidence stayed until she told me, many years later, that her lover had died, his identity still absolutely protected. The conversation languished and I asked no questions.

I believe that only one person was privy to all the complexities of her double relationships, without judgment, without prurience, and without jealousy, and that was Aunt

Marie, to whom she entrusted her private life and her finances and much of my own upbringing, giving her those letters and papers that were not to be willed to the Library of Congress. Marie brought unquestioning love and, having come to college after years in a sanatorium for tuberculosis, a kind of innocence that allowed her to take what Margaret chose and did as the measure of what was right. Marie joined vicariously in Margaret's life, accepting a limited share of her time, and trustfully passed on the knowledge to me in the confidence that what Margaret had done was right and that knowing about it would prove to be right for me. When Margaret needed a place to meet her lover where she could cook and share a meal as the preliminary of lovemaking, Marie went to the movies and turned over her apartment, the same apartment to which Margaret brought me home from the hospital as a newborn, the same apartment at which I dressed for my wedding.

I wish now that my mother had felt able to tell me more about the relationships that were most important in her life for many years, but this was something she may not have felt was in her hands to give, since others were concerned as well. Sorting it out, it has been important to see that what she did tell me was not profoundly at odds with what she believed to be the truth, so that truth comes as a more complex elaboration on the same themes rather than as a dissonance. Indeed, the curious masquerading lie of a lover with whom she had no interests in common stands out as the only real example of a dissonance, off-putting and implausible.

I have asked myself when, as I was growing up, she might have considered talking about some of these things. When I came home after the year of high school I spent in Israel, I came in late one night before leaving for college to find her sitting up in bed worrying about when I was coming in—this after I had been living on my own in another country for a year. For several hours then we sat and talked and I told her about the things that had never gotten into letters, including the various romances. There had been a brief and only partially conscious romance with a young woman there, a relationship whose meaning I was finally able to decipher only when I recognized her pain as the relationship dissolved. That was, if ever, the time when she could have told me that she had loved women as well as men, for I brought her my confidences as a gift, knowing that in hearing of these relationships, she would feel she had me back after the trust she had shown in letting me go off on my own. Still, the individuals I was speaking of were thousands of miles away, and many of them she would never meet, and I was at the end of a chapter. Embedded in her own narrative, she perhaps could not have felt the corresponding degree of freedom, but not long after she gave me a novel, *Dusty Answer* by Rosamond Lehmann, about a young English woman perplexed and enriched by a bisexual series of loves. That night, however, she disconcerted me by turning from my confidences about tenderness and exploration to a discussion of what it would have meant for her professional life if I had been involved in scandal.

In later years we used to discuss relationships in terms that were both explicit and metaphorical and perhaps neither of us was entirely clear which level of language we were using, comparing the obligations of a friendship to those of a marriage, discussing the way in which frictions develop as relationships come to resemble those of kinship, comparing a moment of intellectual revelation to falling in love. When the metaphors ring true, the communication is sufficient. In fact,

Margaret's words are only a concealment to the reader who prefers to repose in a conventional mode of interpretation. Reading *An Anthropologist at Work* or *Blackberry Winter* today, I respond to certain passages with the sense, yes, of course, that is what she is saying. I see now why she quotes Ruth's lines "I shall lie once with beauty, / Breast to breast . . ."[2] or these,

> *We have but this: an hour*
> *When the life-long aimless stepping of our*
>   *feet*
> *Fell into time and measure*
> *Each to the other's tune.*[3]

I read the descriptions of the correspondence between Margaret and Ruth and Edward Sapir, and the poems they wrote to each other, knowing now that at some stage Ruth and Margaret decided that neither of them would choose further intimacy with Sapir, but rather preferred each other. Margaret described to Marie the way in which they talked, sitting and overlooking the Grand Canyon, to reach that decision. The poems that chronicle that triangle are printed together, including my mother's poem "Absolute Benison," published first in 1932, and the implication seems clear.

Absolute Benison[4]

> *Those who delighted feed on difference*
> *Measure the larkspur head-higher than the*
>   *rose,*
> *Can find no benison in burials,*
> *The only absolute that summer knows.*

But those who weary of this variance
Which only an impertinence can name,
Weary of matching petal with pale petal
To find them similar but not the same,

Turn with nostalgia to that darkened
*   garden*
Where all eternal replicas are kept
And the first rose and the first lark song
Since the first springtime have slept.

When Margaret set out to share her life in the form of an autobiography, she intended to bring it up to the present, ending in the late sixties. In the process of writing she discovered the difficulty of telling a coherent story of her life after World War II. *Blackberry Winter* appeared in 1972 with the subtitle *My Earlier Years*, but was not intended to have a sequel. That decision, a decade and a half later, mirrors the concern that was expressed in her 1955 letter, even as her final chapter, commenting on the wealth of important relationships, echoes the earlier phrasing in its title "Gathered Threads." Partly for reasons of discretion and the need to make readers of all kinds feel comfortable drawing on her ideas, and partly because the activities of her later years were so diverse and complex that biographers will labor long to untangle the multiple threads and the different lines of thought and organization, she abandoned the effort in her own account to share the pattern of her later years, to make herself explicable. Instead, she followed only one thread beyond 1939, her experience of being a mother and grandmother, an experience both central and communicable.

These are the roles in which I have known her best, but

inevitably my knowledge of her goes back before my own birth to the years and relationships in which she felt she was formed and to the stories through which she made the present intelligible, with tales of field trips and marriages, like the sea voyages of ancestors whose histories have merged into legend. In writing for the general public, she used her relationship with me and with my daughter as a lens through which to project her complex present, but I find myself continually looking into the past and into other relationships to understand who she was to us.

The words that I write about my mother's relationship with Ruth have a tone of sadness, a mourning cadence, however I try to tune them differently. Some of that sadness must refer to my own experience of being deceived. However I try now to rationalize and empathize with her choices I have to deal with the fact that she did not trust me to do so, and however I try to appreciate the level of honesty implied in her words, it was not complete. Sometimes, however, I want simply to laugh aloud at Margaret's refusal of all forced choices. She kept Ruth's photograph on the mantel and Gregory's on her bureau, and dealt with the ways in which others were inclined to speculate about her private life by sustaining the implication of continuing love for a man who had left her, as if that were the last love of her life. I think it truly was an enduring love, but it was not exclusive nor did she settle for a loveless life thereafter. She said to me once that her besetting sin was her greed for more and richer experience, and she wrote of herself as a gleeful child with a jump rope.

> *And when at last she tore a star*
> *Out of the studded sky,*

God only smiled at one whose glee
Could fling a rope so high.[5]

She would, I think, have liked a life of more sustained and concentrated sharing, and carried a lifelong nostalgia which may in turn have determined the choices I have made for continuity in family life, but there were few indeed who could sustain the demands of more than a fraction of her attention. And so she moved generously through a diversity of relationships, wounded and surprised when others responded with jealousy, and trusting to warmth and zest to make it right.

IX
Sex and Temperament

As I have reflected on my parents and on the geometry of their relationships, I have found myself going back in fascination before my birth to their early meetings and work together, in the fieldwork that seemed so distant during my childhood. At Perry Street we had between two windows one of those curved mirrors that reflect back the whole room, rather than just a fragment, composing it in a circular design that brings out new and disconcerting symmetries. This is what those field trips were, particularly for Margaret, providing paradigms of integrated vision for returning to the fragmented complexities of modern society. As I look back at them, my vision of my parents as I knew them changes and re-forms.

Margaret and Gregory met in 1932, falling in love in an intense fever of conversation and theory building near the shores of the Sepik River in New Guinea, where Margaret had come to work with Reo, her second husband. Gregory, a lean engaging academic bachelor, was already working in the

Iatmul village of Kankanamun when they arrived. She set the process of falling out of love with Reo and into love with Gregory into the context of the anthropological work she was doing and the theoretical questions she was involved with.

When Margaret wrote about that period in *Blackberry Winter*, she emphasized the claustrophobia of the eight-foot-square mosquito room that she and Reo had had built for them in Chambri (spelled Tchambuli at the time), where they settled for a third stage of research, after months of work in two other New Guinea societies, first with the mild, maternal Arapesh and then with the harsh and warlike Mundugumor. They sat with Gregory in that room and talked for hundreds of hours. This was a moment when both personal and intellectual concerns, concerns about gender and genetic endowment, about culture and personality, reached a critical point. The conceptual scheme that they developed seemed to Margaret at the time a synthesis of all the fieldwork she had done to date and a resolution of deep conflicts in her own identity, including the question of how it is possible to be in love with more than one person at the same time if they are different kinds of person. Not only were the three of them gathered in that tiny screened room, but parents and friends and lovers as well, above all Ruth, whose voice was freshly present among them, for her manuscript for *Patterns of Culture* had just arrived. Gregory credited that arrival with the intellectual impetus that moved his own fieldwork into focus.

It is hard to visualize the kind of feverish atmosphere that must have characterized that interval. Under other circumstances, that level of intensity might have led to an immediate affair between Margaret and Gregory, but that seems to me unlikely, given the minimum availability of privacy and Reo's puritanical jealousy, and above all given Margaret's ethic of putting scientific work first and maximizing productivity in the field. An explosion of love and jealousy in that room would have immediately affected their ability to work with the community outside. Instead, all passion was channeled into ideas, and Margaret and Reo telegraphed Boas that they were coming home with major new scientific insights.[1]

Since World War II, it has become fashionable to borrow the anthropologists' term "culture shock" to describe the sense of being overwhelmed by difference when exposed to another culture. The borrowing has been useful in training foreign-service workers and exchange students for participation in societies where there are superficial similarities and profound differences, but the situation of an anthropologist studying a preliterate group that has had minimal contact with Westerners is a different one. Almost every other Westerner who has passed that way has dealt with the contrast not by failing to be sensitive to nuances of cultural difference but by completely refusing to regard the natives as fellow human beings, and so not establishing any but the most minimum conventions of translatability.

In such a situation, a single fieldworker is a barely intelligible alien, striving to establish relationships within the context defined by the other culture, knowing that his or her own thoughts and motivations, referring back to an unknown world, are virtually incommunicable. There is no single moment or period of shock, for the sense of difference is acute from the beginning, underlined by constant contact with decorated or barely clothed bodies, strange smells and foods, the physical adjustment of trying to hunker on the bare ground or sit perfectly straight against a lodgepole.

Within the sense of foreignness one is seeking human contact by discovering layers and layers of subtle and unfolding difference, each layer of difference encapsulating an adaptation to common human problems. Instead of culture shock one should probably speak of this kind of experience in the field as culture stress, continuing and cumulative. It is a stress that leads to a questioning of the very self.

After Samoa, my mother decided never again to go on a long field trip by herself, and she never again participated in the same way as in Samoa, where she was identified by the community as a member of the group of unmarried girls, even though she had the relief of living in a Western household. In Arapesh she had a colleague, but Reo's long journeys into the interior and his scorn for the timorous and nurturant community with which she was left marooned, as hardly having a culture worthy of the name, created some of the same isolation for her. Isolation took a different form in Mundugumor, where Reo felt far more at home and Margaret found she simply disliked the people.

It is not of course unknown for anthropologists to do fieldwork in a society they dislike, since it is not possible to avoid having feelings in a relationship of such intensity, but it is inevitably less productive where there is dislike. My mother frequently answered my question of why she had not worked in this society or that by saying that she felt an incompatibility with the culture. The Mundugumor were fierce and competitive, both men and women rejecting children and embodying what seemed like almost a caricature of Western ideas of masculinity. They had until very recently been a headhunting people, so it was necessary to keep a certain degree of distance and to be constantly on guard. Reo carried a heavy service revolver which was found fifty years later, after Margaret's death, packed in a box of field materials, and turned over to the New York police to dispose of.

Margaret's initial theoretical interest on that trip, beyond the task of doing basic ethnography of undescribed cultures, had been the patterning of personality differences associated with sex roles. Both the Arapesh and the Mundugumor, although they certainly practiced a division of labor between the sexes, seemed to expect men and women to show roughly the same personality type. In Chambri, however, not only was Margaret relieved to be with a group of people she felt to be more compatible than the Mundugumor, she also found more striking personality differences between the sexes, and Gregory was finding such differences among the Iatmul, although Iatmul men and women contrasted in very different ways.

Margaret hypothesized that given a variety of genetically determined temperamental types to work on, cultural conditioning could work toward a single personality type for all individuals or could selectively mold those who happened to be born as males and females so they would conform to two different culturally defined ideal types. Gender—biologically given—supplies a component of temperament, but the imprint of culture is strong enough to reshape a given temperament to make most individuals fit cultural expectations, including cultural ideas of masculinity or femininity. The situation for the Chambri was complicated, however, by cultural disruption, for the Chambri had recently been driven from their home territories by fiercer neighboring groups. Margaret saw among the Chambri a sort of inversion of Western stereotypes, but some of this has been explained since as an adaptation to severe demoralization, men feeling unmanned by the course of events, and women coping.[2] Margaret and Reo had lived within a short period in three sharply differing societies, each exemplifying in a different way the diverse impacts of culture on the individual. They found themselves differently involved and responsive, the contrasts between them as individuals amplified by the contrasting cultures.

Not only were they weary and stressed by a period of very intense work, they also suffered from recurrent bouts of malaria. Nowadays, those who go to work in the tropics start taking malaria suppressants before they leave and maintain them until after their return, and hope to avoid malaria. In the period when Margaret and Gregory did their major fieldwork, there was no way to prevent attacks of malaria, although they could be controlled with quinine so that one was no longer likely to die as the missionaries and colonial administrators of earlier periods often had. The rhythm of fieldwork included a rhythm of recurrent and only partly predictable cycles of high fever, delirium, and heavy doses of medication. Since they knew the course of the fever and knew that it was controllable, the effort was to minimize the time lost from work. Margaret had a bout of malaria that started the day after I was born and dosed herself with quinine, and I remember that bouts continued with decreasing frequency through my childhood, until at last the only reminder of that continuing tie with the tropics was the fact that she could never give blood. When we went to Australia in 1951 and the potent typhoid shots of that time gave me an afternoon of chills and fever, she piled the blankets on me and told me it was like malaria. And once at my father's house I had flu with a high fever and he read aloud to me from Isak Dinesen's *Out of Africa*, so that I hallucinated the creatures of the tropical bush wandering through the bedroom, again equating my experience with the stories they had told me.

In effect, fieldwork under those circumstances involved submitting oneself to a series of altered states of consciousness, where physical stress and fatigue were reinforced by the constant pressure of other minds and other ways of thinking, and the discipline of work, detailed and careful recording, was the basis for sustaining an identity. Such was the setting in which Margaret and Reo encountered Gregory, all three thirsty for conversation. What is extraordinary, in the context of all these difficulties and emotional crosscurrents, is how much ethnographic work they achieved, for even those who have differed with Margaret on points of interpretation of the New Guinea fieldwork respect the quantity and accurate detail of the notes.

In those hours of intense talk, the three of them developed a conceptual model they called the "squares," based on the cardinal directions. This was a system for talking about complementarities of personality produced by the interaction of genetic endowment and cultural conditioning. The fullest statement in print of the squares occurs in *Blackberry Winter*, where some eight thousand words of erratic and fragmented draft were edited down to less than three thousand by Rhoda Metraux—and yet the explanation still remains largely unintelligible because Margaret changed her way of organizing the diagrams halfway through the writing.[3] All this confusion, and the deferral of publication, seem to me to indicate deep doubt and ambivalence about the conceptual scheme itself, which Margaret treated finally, and correctly, as a part of her autobiography rather than of her scientific work.

Both Margaret and Gregory did continue to use the scheme occasionally as a shorthand for characterizing either individuals or cultures all their lives, however, referring to them by the points of the compass. Thus, they would speak of "Northerners" (people like Reo) and "Southerners" (people like Margaret). The compass points were less frequently used for East and West, but the classification was retained: Westerners were called "Turks," managerial people like Diaghilev, the great impresario of the period, while in contrast Easterners, people like Nijinsky, the dancer and artist who went mad, were called "fey."

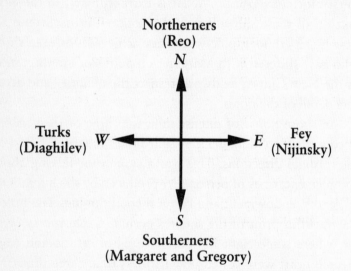

In her draft, Margaret quoted an analogy that Reo used to discuss the four types and also to discuss his relationships with Margaret and with the woman he had been in love with before meeting Margaret, to whom he later returned. "If you go to a dance," he said, "as a man who is a Northerner, you will find that if you want to dance with a girl who is also a Northerner, she will only dance with you if she feels she has chosen you, a girl who is a Westerner will only dance with you a little while because she has been taught she should divide her favors, a girl who is an Easterner will refuse to dance with you for long because she sees herself over against all the others on a stage, but a Southerner, the kind of girl whose shoulders are so rounded that her shoulder straps fall off, will dance with you all evening just because you want her to."[4] It was in this characterization of the Southerner, trying to please a demanding Northerner, that Margaret saw herself in relation to Reo, and in contrast to Reo she saw Gregory as a Southerner as well.

The system was also expressed in terms of two sets of binary oppositions, "possessive" versus "responsive," and "caring" versus "careful."[5] In this phrasing, "caring" indicates concern for one's own feelings, as exemplified by Reo's constant preoccupation with his jealousy of the best-known anthropologist of the time, Malinowski, and indeed of Margaret and Gregory and Ruth. "Careful" indicates concern for the other. When Margaret arrived in Kankanamun, Gregory recognized and responded to her deep fatigue. "You're tired," he said, and for her these were "the first cherishing words" she had heard in months.[6]

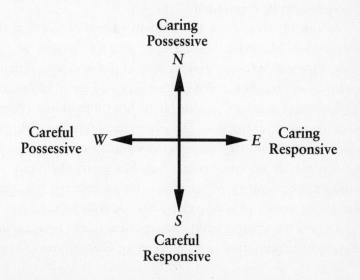

Talking and arguing in that mosquito room, the three
sorted themselves and all the societies they had studied
according to this framework. "Our sense of discovery was
completely combined with our own personal sense of discovering ourselves, Reo as a Northerner, I as a Southerner and
Gregory as a Southerner but a little closer to East than I was.
Vis-à-vis Reo's temperamental interpretations of life, Gregory and I would look as if we were both Southerners, he a
masculine and I a feminine version of one temperament,
which provided the basis for Reo periodically announcing he
was going to go and find a woman who was a Northerner
and all of us speculating about our parents, siblings, friends
and former lovers. As individuals we felt a strong sense of
revelation in finding out who we really were. This played
into my early desire to be certain of my own identity, an identity which would be clear even with many years of some alien
life. I felt as if all the confusions that had been thrust upon
me by other people's conception of what professional
women, or a woman with any brains, must necessarily be,
[were] now to be resolved."[7]

In the squares, I see the kind of pattern seeking that
Gregory and Margaret loved. The scheme set Gregory on to
an exploration of theory making and of the nature of different kinds of models involving symmetry and complementarity. This continued to be central to his thought for thirty
years or more, transposed into a dynamic framework in
which either symmetry or complementarity could produce
equilibrium or disequilibrium. For Margaret, the squares
provided the geometry within which, for complex people, the
alternative worlds of happy-ever-after could be worked out.

In fact, the squares really did represent a step forward in
relation to existing theory. In *Patterns of Culture* and in ear-

lier articles, Ruth Benedict had developed the new and important notion that a culture could be characterized by a single pervasive cultural configuration and the idea of using "personality writ large" to describe it, and Margaret had seen both the Arapesh and the Mundugumor in terms of this kind of common pattern. It was also perhaps this expectation of internal homogeneity that had led her to accept the limited view of their own culture given her by the young girls she worked with in Samoa, seeing Samoan culture as they did, as consistently moderate and serene, without seeking out the more aggressive cultural emphases of the young men. The discussions on the Sepik set the scene for the next important insight, that a culture could incorporate sharply contradictory or contrasting themes, often but not always organized on gender lines, so that by the end of World War II anthropologists could think in terms of layers of complex paradox and Ruth herself could characterize the Japanese in terms of both the "chrysanthemum" and the "sword."

Using the squares, Margaret classified a series of her own most intimate relationships. Her father, an economist at the University of Pennsylvania, a dominant and demanding person, she saw as falling into the same type as Reo, a Northerner. Ruth too was seen in relation to that type, but farther toward the East—a Northeasterner. Indeed, much of Margaret's relationship with Ruth is expressed in this characterization, for most people—myself included—remember Ruth in terms of her Eastern characteristics, as remote and withdrawn, while Margaret was responding also to her capacity for intensity, for jealous passion.

Two of the people who had been most important to Margaret over the years, Luther Cressman and Marie Eichelberger, were seen as Westerners. Margaret and Gregory had in common their Southernness, what we might now call nurturance, and Ruth and Gregory had in common a degree of Easternness, of introversion and withdrawal. Nijinsky and Diaghilev were brought in as conspicuous figures in the artistic life of the period, useful as individuals to exemplify the two types not present there in the mosquito room, and this was especially useful since Margaret, Reo, and Gregory had few friends in common.

It seems to me that the point of revelation for Margaret must have been the possibility of distinguishing clearly between two kinds of difference, both given at birth and both profoundly modified by cultural conditioning. It was highly significant to her that the kinds of temperament she called Northern or Southern could occur in either males or females. Although that difference in temperament can also be used in a given culture to provide exclusive models for masculinity or femininity, as Western culture has tended to demand that the ideal woman be Southern and the ideal man be Northern, and provides a useful match for common aspects of male and female roles, it is not a necessary fact of gender. Thus she felt that her relationship with Gregory was based on similarity— a common temperament—as well as on difference—opposite gender—while her relationship with Ruth was based on an inverse set of similarities and differences.

The scheme still made it possible to think about cultures in which a single temperamental type became the ideal, without a significant emphasis on gender-based contrast, but it suggested a finite geometry within which variations of ethos could be placed, rather than the open-ended set suggested by Ruth's characterizations. It also made it possible to think of cultures in which masculine and feminine roles are culturally counterpointed and contrasted in the elaboration of a polarity that might occur temperamentally in either gender.

As the discussion unfolded, the possible positions in the system were identified with specific societies, one of which, Bali, none of them knew firsthand, but, on the basis of their reading, it seemed to them likely to fit the description given by the framework of complementarities. This was the first step toward the joint fieldwork Margaret and Gregory later did in Bali.

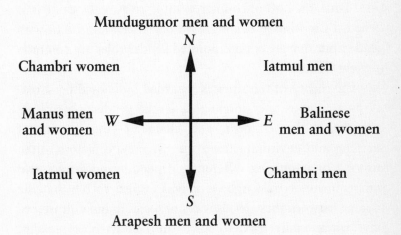

The model also helps in considering the various ways in which one can be a misfit in one's own culture. Ruth Benedict had tried to address the situation of the misfit, speaking from the position of one who had tried to subordinate her own energies and ambitions to cultural ideals of femininity. She discussed a range of possible adaptations, where the person who cannot fit in may simply live in the discomfort of compromise or the stress of uncongenial conformity, or may be given a special role, provided for misfits. It seemed clear that there might be some individuals temperamentally so at odds with their cultures that they could not fit the prevailing style, or the prevailing style for their gender. One strategy for dealing with this would be to amplify the dissonance and turn them into specialists like the *berdaches*, transvestite males in some American Indian societies who could not play the kind of fierce competitive warrior role the society expected of them, but found a social place by adopting a variant version of a female role but not necessarily a changed sexual orientation. Thus, the idea of counterpointing roles really came into Benedict's writing in dealing with deviance rather than with gender, but sharply counterpointed gender roles are common in New Guinea.

The theory of the squares regarded individual personality as a composite of interacting levels: gender, physiologically based temperament, and character—the imprint of learning and environment on the two innate factors, often directed or impinging differently depending on gender and temperament. For Margaret, it was critical to see that the linkage between these levels was not fixed. In our culture, we have traditionally tried to form the character of females, regardless of temperament, to correspond with a temperamental type (Southern) that is one of the range of types that the theory predicts will occur naturally in both genders, and to steer males away from that type.

At some level they must have known that the neatness with which it all fell into place was a kind of intoxication, a malarial madness. Reo repudiated the whole way of working and thinking in a kind of panic, and accused Margaret of simply using it to justify a new romance, while Margaret and Gregory each attempted a partial synthesis. Gregory in *Naven* dealt with polarized (schismogenic) sex roles in the context of a discussion of the unsatisfactory literature available on physical types as a basis for temperament. Margaret explored the three cultures of Arapesh, Chambri, and Mundugumor under the title *Sex and Temperament*, but downplayed her thinking about temperament so that it has been all too easy for readers since to speak of the way she contrasts biological and cultural factors, confounding the different kinds of biological factors with which she is concerned.

Clearly, however, what was pivotal for her was the discovery of a way of sorting out two kinds of biological given, temperament as well as gender, each capable of being elaborated into complex polarities. This allowed her to understand, as well as the ways in which she differed from culturally prescribed patterns, the choices that she made in relationships, without rejecting her own sense of valuing being a woman, valuing her own gender. Margaret never lost her fascination with temperamental givens. This was one reason for her interest in tests and records in the delivery room and she also remained interested all her life in the study of physical types.

There is another reason why the set of ideas that was crystallized in the squares has never been fully and clearly laid out in print. When Margaret and Reo came out of the field in 1933, they came out into a world in which it was increasingly urgent for anthropologists to affirm the potential of all human beings for learning any culturally given model of humanness. This affirmation of commonality of potential, regardless of race or ancestry, had been a central task of anthropologists, a task to which Franz Boas and Ruth Benedict and Margaret over the years devoted themselves. It was made increasingly urgent by the rise of Nazism, has required constant attention in the effort to deal with American racism, and is still urgent in our continuing need to affirm that differences of gender are not differences of inequality. Margaret always remained convinced that her own uniqueness was partially genetic, but for many years she carefully muted her interest in the notion that differences in personality or in intellectual potential might be related to the genetic endowment of individuals, feeling that any effort to deal with these matters would lead to distortions by those who evoke the old, crude dichotomy of nature versus nurture and misuse biological explanations to justify social facts.

Over the years after their divorce, Margaret and Gregory went on meeting in professional contexts and occasionally finding the opportunities for long and rich conversation. One such conversation that I remember brought them into sharp debate on just this point. Gregory contended that it went against any kind of reasonable scientific open-mindedness to assume for a species, such as the human species, with considerable variation in different populations all still capable of interbreeding, that among the different types of variation there were no cognitive differences. Margaret argued with passion that as long as people tend to move so quickly from concepts of diversity to concepts of superiority, and as long as mental variation is treated in terms of such a crude and culturally biased aggregate quantity as I.Q., this question cannot and should not be studied.

In everything that Margaret wrote she took into account the question of what it would be helpful for people to know and how they might move from concept to concept without distortions that would be damaging. She believed in putting knowledge to work and in both seeking and expressing it in ways that would put it at the service of humankind, while

Gregory would move toward the exploration of an abstract idea, reaching for an amoral elegance. Still, when he began to express concern about dangers to the environment, Gregory was arguing precisely that the real dangers come from ideas: from premises, on the one hand, and from conscious purpose and the wish to manipulate through technology on the other. Margaret saw the whole of theory making as having social implications, and knew that whenever she described a world or devised an abstraction, she would be having an effect on how people acted and thus was engaged in social construction.

It is perplexing for me, as their child, to find that Margaret thought both of herself and of Gregory as primarily nurturant, "careful," and responsive, since I tend to see the contrasts between them more sharply than the similarities and to be aware of the ways in which they were less than fully nurturant. Reo saw the scheme as a way of rejecting him, bringing Margaret and Gregory into alliance against him, but it is important to remember that Margaret saw Ruth as a Northerner too—Ruth who remained very close to Margaret even though she protested when she heard that Margaret had formed a new relationship. But in fact Margaret's perceptions and her discussions of the squares seem to underplay the axis on which she and Gregory really differed most, Margaret in many ways a Westerner and Gregory in many ways an Easterner. When they came into conflict it was across that contrast and when I try to understand conflicts in myself between the impulses of poetry and administration, it is also across that contrast.

In later years, Margaret used to quote a poem of Rudyard Kipling's as a way of thinking about this kind of difference. Contemplative Mary of Bethany in the gospel story stays with Christ, while her sister, Martha, busies herself in the kitchen. Margaret would quote his lines in scathing scorn for the way in which the "sons of Mary" leave responsibility to the "sons of Martha" who take it up patiently,

> *They do not preach that their God will*
> *rouse them a little before the nuts*
> *work loose.*
> *They do not teach that His Pity allows*
> *them to drop their job when they*
> *damn-well choose.*[8]

This time the familiar theme of complementarity was expressed with a difference, for Margaret saw that in a given relationship one might be either a "son of Mary" or a "son of Martha." With Marie, Margaret was free to go on Mary's path, to dream and follow her aspirations while Marie took care of the practical details of life, but with Gregory, Margaret was forced into Martha's role—and was resented for it. Over and over, she would involve herself with people whose intellects she admired, who then leaned on her to provide the practical support they needed, but throughout her life she relied on practical support from others.

It's extraordinary to think of her characterization of herself as fitting the description of the "girl whose shoulders are so rounded that her shoulder straps fall off, who will dance with you all evening just because you want her to." Yet it reminds me of all her efforts, in the middle of hustle and bustle and international travel, to be obliging and cherishing— Margaret going out and shopping for particular delicacies for Barkev, planning entertainment for Vanni, holding me on her lap and comforting me when I was upset as a child. The nurturing is deeply there and masked by other kinds of busyness. Margaret arrived in Manila shortly after our newborn, Martin, died, and comforted me with all the traditional forms of succor, embraces and cups of tea, at the same time that she encouraged my analysis of the ways in which Filipinos dealt with bereavement and sat me down at a typewriter to write about it.

Extraordinary too to think of Gregory in this way, for he was often so engrossed in ideas that he was careless of the needs of others, indeed hardly saw them, and yet I can recognize the cherishing moment when his attention was caught, the moment that all his children treasured, puzzled when it was not sustained. Right through my childhood Margaret used to labor to interpret Gregory's behavior into Southern forms. Once I asked her why on earth he had filled a long letter to me—he didn't write more than once or twice a year—full of diagrams of the morphology of some kind of plant, and she replied, "That's the only way he knows to say how much he loves you."

After they left New Guinea, Margaret, Reo, and Gregory traveled home separately, each following the intellectual threads of that sojourn in different directions. The next summer, Margaret and Gregory met in Ireland with another couple, neither couple yet married and both careful to avoid public attention. Wadd, one of Gregory's oldest friends, was for a long time the person Margaret and Gregory looked to for new understandings of how genetic factors enter into behavior, particularly the kinds of behavior we think of as most distinctively human. The intense conversation went on against the cool green of Ireland. Margaret used to describe how, one night, they had arrived at an inn, tired and shabby, and been assigned equally shabby rooms under the eaves. At dinner, Gregory sent back the wine, announcing that it was corked. The host met them as they left the dining room, apologizing profusely that a mistake had been made and shifting them to the bridal suite.

The next time they met was in Singapore in 1935, where they married on the way to Bali, to be sure the ceremony would be valid under British law. Margaret arrived in the Pacific with a trousseau of silk lingerie and Gregory used it to wrap the lenses of his cameras.

X
Parables

The effort to understand the relationship between the ethnographic and the personal for Margaret and Gregory on the Sepik was illuminated for me further when, in 1979, after Margaret's death, I had my single opportunity to go in her footsteps to one of the places where she had done fieldwork, Pere Village in Manus. Manus Island is one of the Admiralty group, off the main New Guinea landmass. Margaret went there on her second field trip when she was newly married to Reo, but it was also the place of her last real field trip. It was to Pere that she returned in 1953 with Ted and Lenore Schwartz and then she visited repeatedly. She was planning a further trip at the time of her final illness.

In 1928, Pere Village (two syllables, like Perry Street) was a cluster of huts built on stilts in the lagoon. It still retained its precontact culture almost intact: Tools were made of stone and bone, the currency was shell money and dog's teeth, and each household was protected by its Sir Ghost, the ghost of the father of the household head, whose skull resided in a bowl hung over the entrance. The work that Reo and Margaret did there typified anthropology as it was practiced then: In the course of six months in the field they attempted to describe a previously undescribed culture, analyzing and learning the language; in addition Margaret was specifically interested in exploring and challenging the assumption that the thought of primitive peoples was like the thought of children—what, she asked, about the thought of primitive children? So remote was the world of the Manus then from the world in which Margaret and Reo lived that when they left Pere the people assumed they had gone into oblivion and would never return.

My invitation to go to Pere came from Barbara and Fred Roll who had spent substantial time in Manus and been part of several of my mother's later trips. Since her death, they had supported a project to build a community center in the village in Margaret's memory. For the opening and the Christmas and New Year festivities, they invited a number of people who had been close to the village or to Margaret's work—Vanni and me, Ted Schwartz, and Rhoda Metraux. The Pere Village that we saw was on the shore, living a way of life reshaped and opened to the outside world. Shortly after Margaret and Reo departed, the village welcomed a Catholic mission to occupy the house that Margaret and Reo had had built for themselves, and accepted Catholicism. World War II brought a Japanese occupation followed by an American invasion that turned Manus Island into one of the principal bases of the war, the staging ground of vast numbers of troops. The village community which in 1928 carried on a way of life that had been stable for centuries was swept up in a deluge of change.

Margaret saw the story of Pere as a remarkable success

story, a parable of hope, for unlike many small and isolated communities, the people of Manus had found ways of adjusting to the impact of the outside world that preserved their sense of identity and dignity. Indeed, she went there hoping to find such a parable and start a new phase of her own life. Losses and disappointments had followed the war: Gregory's departure, Ruth's death, the suspicion and tension of the Cold War and the McCarthy era that made it progressively more difficult to work for social change. She had considered marriage with Geoffrey Gorer and decided that each was too established in a pattern of life, and so she turned back to the field, the well from which she drew her insight.

The story of Pere Village has echoed through my life the way the history of the ancient Israelites echoed in the lives of the Puritans. Ted Schwartz used it at a conference organized by my father in 1968 as a parable of the destruction of an ancient balance. The rebuilding was the model of social change that Margaret drew on for twenty-five years. Going there for the first time in 1979 set much of my life in a new perspective.

The first step for the Manus after World War II came when an apocalyptic religious movement swept across the island, similar to "cargo cults" that have arisen elsewhere in combining traditional and Western ideas. The people of Pere destroyed all their property and threw away their canoes and fishing nets and whatever Western money they had earned during the American occupation, in anticipation that the ghosts of their ancestors would come in ships, bringing a cargo of Western wealth for the people of New Guinea. Traditional customs, membership in the Catholic Church, and much of the economic base of life was destroyed. It was on the burned-over wreckage that remained when the hopes of the cultists were disappointed that a new leader emerged, a man named Paliau. The cult leader was dead; he had instructed his followers to kill him so that he could go and find out from the ghosts why they were delaying.

Paliau had many ideas about the modernization of Manus life, about economic change and education, about democracy and family structure, based on observing the Australians and Americans. Most important was the imperative of moving from the old houses on stilts in the lagoon onto the shore and working together with the shore people, the Usiai. He also restored a rudimentary church, based on a small number of texts and forms retained from the Catholics and other mission groups, but he seemed in that context to be a predominantly secular leader.

Thus, when Margaret returned to New Guinea after a quarter century, it was to find that the people she had first studied in 1928, then still in the Stone Age, had crossed over and now felt themselves to be part of the modern world. Reentering the village on her first field trip after many years, she found that her role was completely changed. Instead of being an alien from a separate sphere, she was now seen as a consultant for the people of Pere who could draw on her knowledge of the wider world and of their own past to build the type of community intelligible to both, according to shared values. She became very much a partisan of the social experiment under way and used it to emphasize the advantage of adopting a single embracing and consistent vision, the vision of a new way that Paliau had presented. If a people who have dressed in grass skirts and shells decide to borrow Western dress, she said, they must also borrow soap and needles and thread, or cotton clothing will soon be turned to rags. It was only after she left, while Ted stayed on, that he became aware of the continuing inconsistencies, the ghosts still moving underneath the surface.

The village stretches along the shore of the lagoon, sheltered by a coral reef. Barbara and Fred Roll had had a house built for themselves at the center of the village, raised on piles and thatched with pandanus palms, so much of the life of the village passed before our eyes. To the south side, we could look out to sea and watch the outrigger canoes passing on their way to fish or sailing to Lorengau, the provincial capital. Here the Southern Cross spread at night, low over the horizon, and in the daytime we could watch the children in the water, sporting endlessly with their child-sized canoes, swimming and surfing over the reef.

Although most of the Manus now live on land, they still cling to the shore, building only privies and pigsties out over the water where the houses used to be. The sea still cleanses the community. From the Rolls' house, a thin catwalk of boards leads out to a high stilted privy, past the pen of a pig who eats the household scraps from a single giant clam shell. Unlike those of the other buildings over the water, the catwalk of this privy has a railing, for Margaret broke an ankle in old Pere Village in 1928 and I was constantly conscious, as people helped me in and out of canoes, of their expectation of physical incompetence. The walkway to the privy was the best place from which to watch the sunset, not quite visible from shore, shockingly beautiful across the water beyond the silhouettes of other islands on the horizon.

On the north side of the house passes the main thoroughfare of the village, broadening into a square which is used for all major gatherings. The new concrete community center is opposite. Just to the east stands a huge tropical tree. At first dawn, children scavenge there for fruits that have

fallen down during the night, and one or two huge fallen
blossoms lie beside the path, white petals surrounding a
shock of magenta-tipped sepals. In the daytime its shade is a
favorite gathering place where women spread out fruits or
vegetables for a tiny market or village meetings are called.

This house, I realized, although better than any field res-
idence Margaret ever had built because it was designed for
long-term use, meets her two most important requirements—
a vantage point from which it is possible to see what is going
on and a place in which life can be organized to allow the
hours of work that must be sustained parallel to participation in the community. If you go and live in a single native
household, assimilated to the role of daughter or sister, you
are immediately constrained by that role—you cannot behave in ways that would embarrass an adopted brother or
loiter on the path with the political rivals of your "father."
Your time may be eaten up by the activities appropriate to
your role and there is no space in the day or in the household
for typing notes or developing film. There is a sense of
achievement that goes with being taken in, treated as one of
the family, but finally you can only report what it feels like to
be an adopted daughter or son, not how the whole fits
together. Margaret in the field spent longer hours typing and
cross-indexing notes, looking for gaps and parallelisms, than
she spent observing, for when she was observing she would
record in dense detail, full of abbreviations, her pencil flying
illegibly across the page, producing a record that would have
been useful only to her if left in that form, something private
rather than the scientific record that sits today in the Library
of Congress for other anthropologists to use.

Margaret looked at fieldwork from the point of view of
how much work could be done in a limited time. I never

accompanied her to the field, but in 1974 she came with Geoffrey Gorer to visit my household in Tehran and we traveled across the country, alternating visits to tourist sites with visits to other anthropologists, who were usually hungry for the chance to discuss methods and share their insights. In one place, our hosts took us up to the flat roof of their house, pointing out that they could see down into the courtyards of their neighbors, solving the problem of the high walls that make Iranian fieldwork so difficult. This she approved at the same time that she wondered why so much time had been spent on duplicating an Iranian-style household, a great deal of extra labor for a small gain in empathy.

Margaret believed that every moment is precious in the field. She avoided paying people to talk about their culture, treating them instead as colleagues with a common interest, but she was not hesitant about other economic relationships: paying people to build a house or hiring them to do the tedious tasks of the household. As a museum anthropologist, she often purchased artifacts for museum collections, and in Bali, where the traditional system allows for patrons of the arts, Margaret and Gregory would pay for performances and ask to have them by daylight for photography. Margaret and Reo amazed Gregory by their sense of purpose and energy in the field, by the fact that they went out and got information, pressing people for answers to questions instead of waiting for chance occurrences.

The days we spent in Pere gave me the opportunity to meet a community I had heard about all my life, just as they had seen pictures of me that my mother put up on the wall of her workroom. We mourned Margaret again as we were led at night onto the shore from our canoe, falling into the arms of the village women and lying with them on the ground in ritual weeping. She was remembered by them with great love, these strangers, as central to their life in a different way as she was to mine. They beat the mourning drums at her death and they sang for me the songs they had composed, which Ted Schwartz translated from the Manus language,

> *Makaret our mother,*
> *Mother to us, your sons and daughters*
> *from a later time,*
> *This side of Papua New Guinea is where*
> *the sun sets.*
> *You have come and left these people to us*
> *as your legacy.*
> *They say Pere is your village and brother*
> *to New York.*
> *Makaret our mother,*
> *Makaret yaye yoya*
> *Makaret yaye yoya.*

To find many of the men and women I had heard about over the years, we had to visit the graveyard down along the shore. But among the living, our strongest link with my mother's times in the village was through the family of John Kilipak, or J.K., who had worked in her household when she first visited. He is an old man now, almost the last of the little boys that she trained to do the housework in 1928 because they were at that age quick and responsive whereas those who had reached adulthood were sullen and inflexible, imprisoned by responsibilities and obligations. J.K. has great warmth and dignity, and yet has always been somewhat at the edge of things, neither completely a part of the new nor the old, naturally agnostic and skeptical. His wife, Siska, was one of the first to greet us when we arrived, bringing Vanni and me the first of many scores of necklaces of shells and seeds.

As the days passed and the actuality of Pere began to merge with my imagined picture, Ted pointed out how, faced by the frustrations of social and economic change, the people of Pere were gradually reaffirming traditions of the past they had abandoned. I found my attention moving back and forth, watching the unstable marriage of the old, the borrowed, and the new.

On the day of the opening of the Margaret Mead Community Center, the women gathered to dance, wearing bras and grass skirts over shorts or knickers. Then Siska came walking proudly through the village with bare breasts, wound with necklaces of bead and shell, with red and yellow croton leaves inserted in her armbands and croton branches like feather fans in her hands, and those who had been faint-hearted went home and put away their bras. We went out to the reef in a canoe so that we could join the official visitors from the provincial government in a ceremonial arrival. Women stood dancing on a line of canoes perpendicular to the shore, making a channel through which the boats of visitors could pass, while on the shore the big slit gongs were beaten.

Then all the visitors were seated in the shade of a makeshift awning and dancing continued. At first, only a few men were involved and they were dancing like the women, wearing grass skirts and following their steps, but then several men came out with spears and more warlike postures; finally, they went up and danced on painted poles, the trunks of trees, mounted two or three feet above the ground. For the first time since the village moved to the shore and into the forms of modernity, first one male dancer and then another appeared, wearing only the white cowrie phallic shells, balancing and leaping on the poles and flipping their penises, weighted and elongated by the shell, back and forth against their loins. Just as at a Fourth of July pageant, the celebration of the present included an evocation of the past.

The history of Pere, with all its layers of change, and of my mother's visits there, was also traced inside the Margaret Mead Community Center in a display of photographs the Rolls had assembled. The earliest ones were taken in 1928 and show Margaret working with the children, looking almost like a child herself; people remarked on how like me she looked in those early years. Then there were pictures of the adults of that time, dancing as they were today, making the great ceremonial exchanges of wealth that were the central preoccupation in the life of the Manus. These exchanges were supposed to end with the reform, but have gradually drifted back. Later pictures show the village established on the shore: men in shorts and shirts, school, drills, committee meetings, and Paliau on a visit, campaigning for the House of Assembly, the Papua New Guinea Parliament. Most recent were the photographs taken in 1979 when John Kilipak came to the United States to work on a dictionary of the Manus language with Ted Schwartz and visited New York to present a gift from the village to the American Museum of Natural History in memory of Margaret, and to see the diorama of the old village in the lagoon, the re-creation of his past.

Just as the village is mixed today in its attitudes toward past and present, so Paliau today is less clearly a heroic figure looking into the future, the charismatic leader I was told of as an adolescent. The people of the village are no longer united in supporting him, and the premier of Manus, who came for the opening, spoke of him coolly as primarily the leader of a sect. Shortly before our visit, Paliau had announced major shifts in the theology of his church, a new version which he said he had always known but had not dared to reveal. Instead of God, the faithful were to name and worship Wing, with an evocation of flight and the great wings of heaven, his son Wang who may also be called Jesus, and Wong, who combines the angels, always important in Paliau's theology, with a reminiscence of the Holy Spirit. All of the missions are to be expelled, the Bible rejected, and Manus and then New Guinea are to be united behind Paliau's leadership. The various efforts at development in recent years and the granting of independence to Papua New Guinea have meant that many of Paliau's secular ideas became the common sense of a new era—common sense accepted by a generation that has gone to school and includes followers of all the principal missions on Manus, as well as increasing numbers whose relationship with the wider world has never been mediated by any kind of religious symbolism. At this stage, the emphasis of the earlier period on joining the modern world and taking a fair share in it becomes canceled out by Paliau's need to differentiate and contrast his own message with that of other groups, so has suddenly revealed a revision of his message that constitutes a startling localization.

My mother had considerable sympathy for the church established in the village after the departure of the mission and the ravages of the cult. When she came back from the 1953 trip, one of her early concerns was to send candles to the village, and she sent them a reproduction of a painting of God the Father brooding above His son on the cross, which is still set on the altar during services, although mildew has crept in under the glass. She felt that in a world of rapid change and increasing intercommunication it was wrong and sentimental to expect people to remain marooned in their precontact beliefs which neither provide a coherent system of meaning for confronting a wider world nor offer membership in a larger community. Because she valued religious forms of expression, she would have liked to see the church in Pere reconciled with the Catholic Church or once again in touch with some part of the Christian community that might, with forbearance, give them access to a broader fellowship. But the church today is reduced to a handful of members, its liturgy a falling echo.

Still, I wanted to join my sympathy with hers, so on both Sundays in Pere and on Christmas I put on shoes and clean clothes and went and joined in with the services. This I could do because the services are always conducted in the lingua franca of Papua New Guinea, which bridges its hundreds of indigenous languages, Pidgin English or Neo-Melanesian, even though the congregation is linguistically homogeneous and it would be possible to use the Manus language, *tok ples*, "talk of the place."

For Paliau, Pidgin was a symbol of unity, unity between the sea-living Manus fishermen and the people of the interior, and unity of the different islands and language groups. Just as Pidgin allows communication among people from different areas, it gave me a means of entry. As a language it is splendidly serviceable, using limited resources of largely English vocabulary flexibly and resourcefully within a grammar that is basically Melanesian. One of the problems of Westerners who try to learn it is that, even as they recognize it as a product of culture contact, they fail to perceive that it is also a real language with its own structure and its own semantics, although this perception is helped by the writing system which is strictly phonetic.

My mother used to speak Pidgin at home occasionally—she had certain phrases she used routinely, like *Samting bilong yu yet*, which means something like "That's your problem." Others she used in anglicized pronunciation, so I never realized they were Pidgin until I went to Manus, like *something-nothing* and *rubbish*, her favorite term of disdain. She could also occasionally be cajoled into speaking Pidgin or telling a version of Genesis 1 beside the Cloverly fireplace. *"Nau long taim bipo, i no gat wanpela samting i stap. I no gat man, i no gat meri, i no gat pik, i no gat bulmakau, i no gat wanpela samting tasol. Nau wanpela polisman i stap on tap, nem bilong em i God. Nau God i tok, mi tink mi mekim olgeta samting. . . ."*

As a child, I parsed this by associating each element with the English word or fragment from which it was derived: *Now, long, time, before, he, no, got, one, fellow, something, he, stop.* Properly speaking, however, it should be translated, since many elements derived from English are used in different ways: "A long time ago, there was nothing in existence: no man, no woman, no pig, no cow, there was nothing at all." It is easy to accustom your ear to the *i* (derived from *he* and pronounced to rhyme with it) and see that this occurs with all third person verbs, whether the subject is stated or not, that the *-im* (from *him*) is tacked onto transitive verbs, whether the object is stated or not, and that *-pela* (from *fellow*) identifies adjectives. It is much harder to decide how to translate *polisman* . . . policeman? authority figure? person in charge?

I remember the zest with which Margaret spoke Pidgin,

giving an effect that was both comical and racy. Gregory also had his own standard recital, the story of the monkey and the pussycat, both animals not native to New Guinea, who steal two white children and gradually acquire for them the accouterments of a European life-style. In Pere, I found that it was much harder to understand Pidgin when it was spoken by Melanesians, but because I had read analyses of Pidgin as a graduate student in linguistics and used Pidgin texts with classes, it gave me a beginning mode of communication, albeit one that might always be a source of distortion.

Pidgin has none of the echoes of myth and poetry that enrich liturgical language but instead has captured within it, like flies in amber, the attitudes of European prospectors and plantation owners who needed a language for giving orders and assumed that words like *savi* ("know"—from the French) and *pikinini* (which is based on Portuguese) would be more easily intelligible to natives. An outsider has to try to separate the words in their present meaning from their etymologies: *Bagarapim* simply means "destroy" or "spoil" (*bugger up*); *pikinini* refers to any offspring, black or white, including fruit of a tree.

The people of Pere sang Christmas hymns:

> *Mi lukim yu, O Yesus King.*
> *Yu pulim daun yu boi nating.*
> *Bikpela God, O Yesus yu,*
> *Mi lukim yu, nau mangki tru,*
> *Long Bethlehem, long Bethlehem.*

The words struck bluntly on my ears. Barely able to comprehend, I stood in the bare, lighted community center struggling to escape the echo of white colonials calling the naked village children "monkeys" to a simple word now meaning "little boy" and the shared image of the Christ Child. Only gradually was I able to smooth the periods, interpreting *bikpela God* as "Almighty God," and *boi nating* as "a mere servant."

To fill the three-hour service that began at 3 A.M., it was necessary to use three times over all the hymns and carols and all the prayers in the booklet that did not refer specifically to other festivals—including the barely recognizable parts of a mass separated from its actions and from any sharing in bread and wine, marked off as special only by the rising of the congregation for those passages. The borrowed forms are as skimpy as the rags of Western clothes when these no longer are replaced or cared for. A Star of Bethlehem made of cardboard, with balloons and a lighted flashlight hanging from it, had been rigged with a pulley at the back of the community center, to be drawn up to the front of the room during a hymn, above a hutch of branches and palms built to represent the stable, which stood empty, for the community had no figures to set up a nativity scene.

When a system becomes so impoverished, it is likely to continue to decay unless there is some new injection of vitality. Aside from all other possible reasons for attrition, communities that have gone through periods of change and innovation become subject to boredom. In a stable ancient system, such as the one that existed in Pere in 1928, boredom is not a significant subjective dimension of experience. Monotony and repetition are characteristic of many parts of life, but these do not become sources of conscious discomfort until novelty and entertainment are built up as positive experiences.

In contrast to this, Ted has used the term "meta-stable"

to describe the state of the Manus after the cult, a state not unlike our own in which change is itself valued and experiences cannot be repeated because their original novelty was so great a part of their meaning to people, so that, for instance, Americans have begun to find even space launchings dull and repetitive. In villages the world over, repetitive rituals draw their value from the fact that they are reliable reaffirmations, bringing people together within the daily round, but it seems clear that simple boredom is a part of the slippage of the Paliau movement and a reason why Paliau himself is trying to inject novelty. More serious is the fact that the tradition is not woven into a complex interlocking web of associations, but has a curious flat quality, like words spoken in a room where there are no echoes. Margaret's interest in supplying a picture and candles for the altar comes in this context, against the background of her delight in high-church ritual and her use of a literary style in which every word was evocative, calling up echoes of myth and poetry and liturgy.

In 1928, there was no way in which Margaret and Reo would have felt it appropriate to promote change or take sides in the decision making of the village, although inevitably they were aware that certain ideas were planted by example and consciousness increased by questions asked. Even the simple medical care that anthropologists give in the field when sores and sickness are shown to them must have an effect eventually. In 1953, however, change had produced a new order of self-consciousness, and the people of Pere could regard Margaret as a colleague in the business of social construction, a colleague with wide comparative experience to be consulted, instead of someone from a world so unknown that it was seen as irrelevant. Once when Margaret was in Pere in the late sixties, the village leaders had come to consult her about the problem of children who had finished elementary school but were denied the opportunity to study further. At that time, Margaret proposed that this new group of young men organize as a service corps for the development of the village, describing such projects as the Peace Corps and the National Service Corps she had supported in the United States to replace the draft.

The process of planning and consulting continued during our visit. On the day after Christmas, it poured with rain all day long. Three of the most highly educated young men of the village, who are now working elsewhere but return to advise, came to discuss the development prospects of the village with Fred and Barbara. This was a chance to hear the practical side of the disillusionment of the village with the *nupela pasin*, the new way, a catalogue of projects that have failed, partly because of the way in which a communal ideology was promoted early on, partly because of the pressures on individuals created by the revival of the old gift-exchange system.

A common smokehouse was built at one time, but not maintained; a copra plantation was offered to the village if they would develop it but interest waned and it lapsed back to the government; a freezer for fish that the young men working outside had given the village has never been effectively used. Now the village dreams of buying a truck to run a transport service near the provincial capital, and Ted and the Rolls sat discussing with the young men the need for market research, the need for accounting and maintenance, the problem of rising prices of gasoline . . . ending up emphasizing that the village will probably be most successful in the area of their past experience, fishing.

I sat on the sidelines, listening to the stories of problems in the microcosm and reflecting that the issues were as much issues of belief as were the issues affecting the church. The terms were not used, but we were clearly discussing the waning of an ideology and the reappearance of the old individualism moderated by the obligations of kinship, while offering and considering the questions of free enterprise and the ethics of good management.

In fact, fishing continues to be the main resource of the Manus. Margaret felt that their physical skill, exhibited above all in canoeing and fishing, and the way in which confidence in these skills was developed in a free, autonomous childhood, was one of the things that enabled them to retain their poise and dignity through a major cultural change. The ideology of the *nupela pasin* did not lead to an immediate abandonment of the old-fashioned ways of making a living, nor did the Manus undertake cultivation and the shore people learn to fish, as Paliau had urged.

A few days later, Karel Matawai took us out fishing, accompanied by two other men, in his canoe with a powerful outboard motor partly funded by the Rolls. Manus canoes are based on the dugout trunk of a large tree, connected by transverse members to an outrigger, with a platform built over the water on the cross pieces, to which an outboard motor can be fastened. In the old days, the canoes had sails as well as poles and paddles, originally woven mats, but today almost no one relies on the wind. Looking at the canoe, I was struck by it as a symbol for the way in which different cultural elements have been combined. Steel tools are used in making it, and occasionally bits of wire have been used for mending, but basically the canoe preserves its traditional form and construction, including the binding of the transverse poles with rattan to supports on the outrigger made of many thin rods, which allow wind and water to pass through freely. Now, in an effort to preserve these skills, the children make scale models of canoes in school, and Vanni took one home to present to her school in Cambridge. The canoes are no longer propelled by wind, however, but by an imported gasoline engine, and one wonders how long the skills of sailing will remain available.

It now looks as if, for this community, the point of no return may not have been passed when the old skills are needed. Karel took the canoe out and then quartered back and forth on the Bismarck Sea, looking with an ancient skill for seabirds swooping over the water and betraying a school of fish. Again and again, he would adjust his course, pointing the canoe in a new direction, and then my eyes would see the birds long after his, as if the pointing of the canoe had brought them into being. Then we would head through the school of fish, the three men trolling with their nylon lines. One, two fish would bite, and then we would be searching again for birds to tell us where the school had re-formed, to cut again across the water. Each fish was dropped down into the hull of the dugout where it flopped and struggled in the moist darkness under the platform. In a morning's fishing, two and a half gallons of gasoline were used and only twelve fish caught, but on other days and with different luck the catch might be ten times as much. In the old days the men would have worried about the possible casting of black spells on their fishing equipment after such a poor catch and appealed to their diviners and Sir Ghosts.

A large number of outboard motors have been acquired at different times by the people of Pere Village—the Rolls and Ted once counted seventy, only ten of which were still operating, for mechanics and maintenance skills are not available. As people perceive it, the economics of fishing does not really include the cost of the motors, for such capital expenditures tend to be made with the wages of the educated young people who are working outside the village, all over New Guinea, many as civil servants, so that most of their wages come from foreign aid. The rising cost of fuel, a daily running expense, will begin to be compared with the value of the catch, however, and availability may be a problem, especially since the distribution of fuel to remote villages is itself expensive.

In 1953, it did not seem unreasonable to hope that working within new attitudes and social institutions the Manus could develop their communities and use their skills with economic effectiveness, so that the material base of their lives would be radically changed. Those were the years of the United Nations Development Decade, when it was expected that the whole world would share not only the symbolic systems but also the economic prosperity of the West. Watching the fishermen with a skepticism about economic development that I learned in the seventies, underlined by Gregory, it seemed to me that Paliau's modernization program was as ill-founded in its economic assumptions as the most bizarre forms of the cargo cults. It is no more likely that the earth will support ever-increasing consumption for ever more people than that the ghosts of the dead will deliver that wealth directly to the doorsteps of their descendants.

My mother was fascinated by the notion that there would be enough so that everyone would have a share; the hope for a chicken in every pot echoed across the Depression, and when chicken-in-parts was invented she felt that human ingenuity would be able to give each child a drumstick and each person a share in prosperity. Neither in Pere nor in the developing world in general have all her hopes been met. And yet . . . and yet . . . in spite of economic problems, Pere village is more vital today than thousands of other villages that have gradually become dependent on external prosperity, a community of people with a continuing sense of their own competence and of the capacity to make choices. Margaret and Reo had described the old-fashioned Pere as a harsh, glum place, in spite of its tropical island beauty, while today it is warmer and more relaxed, with men and women working companionably together and individuals released from their traditional rigid roles and burdensome obligations.

Again and again, I found myself reflecting on my mother's interpretation of the story of this village and of the Paliau movement and the burden they were made to carry in understanding social change. My mother's picture was too clear and optimistic for the reality and yet the reality is impressive. As so often when I look over her work today, it seems to me that she was both right and wrong, above all right in knowing that there is a range within which communities can choose their future. Such choices will always be mixed and success will never be complete, but the affirmation of the possible is surely the place to begin.

On the day before our departure, Paliau himself came to visit. Because of the division in the village, Paliau was not allowed to use the community center, and in the end his meetings were held outdoors. He arrived in the morning in white Western clothing, and spent an hour or two going over his new theology with Ted, surprised that Ted found no difficulty with the complex notion that Wing sees no evil, Wang hears no evil, and Wong speaks no evil. Then his supporters set up a table by the Rolls' house, in the shade of the baobab tree, where he displayed photographs of himself and campaign material and his Order of the British Empire.

Curiously diminished from the figure I had heard about for so long, he stood to speak in front of the community center while a troop of his supporters marched in from the east end of the village, singing hymns composed for the new theology, to stand at attention in the sun, spaced out in drill formation while he spoke about the history of the *nupela pasin*, his conversations with my mother, and his newest religious ideas. Those who did not think of themselves as his followers were still present to watch, listening and sitting comfortably in the shade around the edges.

In the afternoon, Paliau and his wife, Teresia, who had been resting in one of the houses, emerged in traditional costumes, both of them notably lacking conviction and Teresia especially bored and embarrassed. Then, as darkness fell, a meeting began to gather beneath the massive, ancient tree, while we listened from the veranda. The form was parliamentary and a chair was elected, but the purpose of the meeting was ambiguous, as was Paliau's role. Some felt that this was an occasion for regular civic business of the village, but Selan, J.K.'s son, who had murdered his wife in a psychotic episode some years before and now occupies himself with a clowning club and with religion, stood up and said the village should unite in deciding whether they believe in the "study group" or the "red book," representing the old and new versions of the Paliau church.

Peranis Tanno, who had been a leader in Pere and in Manus generally for many years, stood and made a bitter speech at Paliau, reviewing the support given by the people of Pere in the early years of the Paliau movement, and berating him for not attending the opening of the community center, saying that Pere is being made "rubbish." It had been four years since Paliau's last visit to the village, and in that time his continuing support had dwindled to less than a third of the community whose life still bears the stamp of his thinking. Speech followed speech, with Paliau rejecting criticism until some proposed adjournment.

In the shadow of the great tree, the meeting had slowly been covered by darkness, although a moon was shining over the water. Suddenly, Selan leaped to his feet, denouncing the meeting and the village, denouncing the church, denouncing the Rolls, his voice suddenly harsh and tense and insane, filling the darkness with threats and menaces. A man in the crowd shouted, "Come on, Selan, how about the Funny Group?" and others joined in and somehow he was cajoled into pulling out his harmonica as someone stood up and did a comic dance for him. Once more he switched into the mode of threat and denunciation and once more he was coaxed into playing the clown, and then a group of men got him back to his house and set a guard for the night as the rest of the villagers scattered, knowing that once in the past Selan had killed.

The next day we departed from this village which still balances precariously at the edge of the modern world, so small a place to have been used as a paradigm for change in discussions and articles read by so many, but a place that could sing, "They say Pere is your village and brother to New York."

XI

Participant Observers

Both Gregory and Margaret combined their efforts at understanding biological and social process with the effort of self-
knowledge. Each drew on a diversity of ways of knowing
and thought about knowing as a central question; each,
intrigued by the nature of learning, brought special care and
elegance to the act of teaching. It is not accidental that when
Margaret was on the Sepik, struggling with the question of
diversity in herself and in her ways of loving, she was formu-
lating the contrasts between three New Guinea peoples who
dealt very differently with maleness and femaleness, with
assertion and creativity. And it is not accidental that Margaret's excitement about the new beginning made by the
Manus after World War II echoed her decision to make a new
effort at fieldwork herself and to draw the connections
between the prewar studies of the preliterate communities
and the wartime studies of complex modern societies, conducted without the intimate contact of fieldwork. She too
was making a new beginning.

For Gregory also there was a continuing relationship between personal and professional understanding, always emphasized by his search for abstract models so that formal similarities could be examined. After the war Gregory was involved with psychiatry both as a patient and then in ongoing research on the therapeutic process. The same decade that began with his rebellion against Margaret, a rebellion shot through with resentment against his family and especially against his mother, ended with an analysis of patterns of communication in the families of schizophrenics, above all of the role of the mother.

These resonances between the personal and the professional are the source of both insight and error. You avoid mistakes and distortions not so much by trying to build a wall between the observer and the observed as by observing the observer—observing yourself—as well, and bringing the personal issues into consciousness. You can do some of that at the time of the work and more in retrospect. You dream, you imagine, you superimpose and compare images, you allow yourself to feel and then try to put what you feel into words. Then you look at the record to understand the way in which observation and interpretation have been affected by personal factors, to know the characteristics of any instrument of observation that make it possible to look through it but that also introduce a degree of distortion in that looking. All light is refracted in the mind. To look through such a lens, it becomes important to know the properties of the lens. This is the scientific goal of biographical work on social scientists.

In anthropology the relationship between observer and observed is complicated by the fact that one is constantly moving between two conflicting impulses, an impulse of closeness and an impulse of distance, the desire to leave home and the desire to discover oneself at the end of the journey, to go away to worlds rich and strange and to discover in them the ordinary, recording and explaining what initially seems exotic. Anthropologists are not, finally, romantics. Occasionally, as in Audubon's paintings of birds, the impulse to capture in precise description yields an object of beauty, but it is a different impulse.

Gregory's decision to become an anthropologist was both an affirmation of family tradition and a rebellion, a way of leaving home which, in taking him to New Guinea, was more than simply geographical. He shifted into anthropology from biology and found a field with very little in the way of analytic models, all of which left out a great deal: You could begin to describe social organization; you could look at the way different customs served the needs of subsistence and adaptation; and you could collect details of myth and custom, sorting and comparing. At Cambridge, anthropology was still very much a branch of natural history, however, with the emphasis on description. For forty years, Gregory struggled with the question of how to provide anthropology with the clear, taut framework of fundamental ideas that would make it truly a science.

My own sense of Gregory as a scientist was also rooted in natural history, the way of doing science that he had learned as a child. His father, William Bateson, was a distinguished geneticist. He expected that all his sons would revere the arts but make their contribution in the sciences. This entire burden of familial expectation fell on Gregory after the death of his two brothers. The oldest, John, was killed in World War I; this loss was followed by the shockingly public suicide of his middle brother, Martin, on John's birthday. The two older boys had initiated Gregory into natural history as

Gregory later and fragmentarily initiated me, and John wrote from the trenches before he was killed about the birds and insects that somehow survived to fly across the battlefield. Martin, while still at school, had told his father that he did not believe he would be a scientist and had begun to experiment with writing poetry and drama, only to encounter parental discouragement and skepticism. He had written to his mother, "My highest ambition is to write poetry. But I know I cannot unless some of what I have already written is. I could be a humble follower of science; but if I write poetry, I must be a leader. . . ."[1] After Martin's death, William wrote a long letter to Gregory urging him to find comfort in dedication to scientific work:

> The faith in great work is the nearest to religion
> that I have ever got and it supplies what religious
> people get from superstition. . . . Of course,
> there is great work that is not science, great art,
> for instance is perhaps greater still, but that is for
> the rarest and is scarcely in the reach of people
> like ourselves. Science I am certain comes next
> and that is well within our reach, at least I am
> sure it is within yours. It was just because I could
> never see that Martin had the real spark of art
> that his change of plans was alarming.[2]

There is a curious resonance between this correspondence and Margaret's own reveries about writing poetry, which she came to regard as no more than an avocation because of her college friendship with the poet Leonie Adams: "Without her I might have gone on much longer, fancying that a slight talent was a real gift."[3] "It seemed to me then—as it still does—that science is an activity in which there is room for many degrees, as well as many kinds of gift-edness."[4] In this way, science contrasted with art for her as well.

When Gregory decided to shift to anthropology, he described that shift as a shift from "ordinary impersonal science" to "a branch of science which is personal where I should be able to take root a bit," one that would provide the sense of "personal inspiration" he felt he lacked.[5] His letter almost seems to hint that anthropology was a kind of compromise position which would resolve the opposition between science and art. The themes of art and science are curiously mixed in all his work.

Anthropology is probably the most personal of the social sciences, for the circumstances of research are often such that it is impossible to divide space and time, shifting gears from a personal to an impersonal mode and working within a for-mally structured framework of attention. Both Margaret and Gregory developed a style that involved collecting observa-tional material in the expectation that, however rich and bewildering it might seem at first, they would arrive at points of recognition when things would "make sense" and fall into place. In the search for such moments of insight they would be dealing with points of congruence within the culture they were looking at and also points of personal response.

The process is an aesthetic one, one of listening for reso-nance between the inner and the outer, an echo that brings the attention into focus. Poets work this way as the curve of a leaf evokes the poignancy of a past moment. Therapists work this way, moving back and forth between their own task of self-knowledge and the task of understanding a patient, knowing that without double insight there will be no insight at all. Indeed, I have always thought of this effort to become aware of and draw systematically on internal processes in the terms of Erik Erikson's description of clinical method as "disciplined subjectivity."[6]

In anthropology, you usually cannot specify in advance what it will be important to pay attention to. This was particularly true in the twenties and thirties when anthropologists had few theoretical models and were often working in previously undescribed societies. One must be open to the data, to the possibility that very small clues will prove to be critical and that accident will provide pivotal insights. You go out, ready at least to do natural history, as the three Bateson brothers went out in Gregory's boyhood, with their butterfly nets and collecting boxes. You may take a few specific collector's ambitions but you will be attentive at the same time to whatever you see and ready to find something quite unexpected, like a rare beetle that little Gregory found in the hollow of an old elm tree, a beetle rare enough for John to publish a report of its finding.[7]

In contrast, for many kinds of psychological research, the observer's attention is very highly specified and he will record only certain defined types of events. Similarly, a sociologist's data are often structured by pre-set questions. But in anthropological fieldwork, even when you take with you certain questions you want answered or certain expectations about how a society functions, you must be willing to turn your attention from one focus to another, depending on what you are offered by events, looking for clues to pattern and not knowing what will prove to be important or how your own attention and responsiveness have been shaped. Margaret argued for years against the effort of granting agencies to shape anthropological research to fit laboratory models of hypothesis testing, with the narrow and structured attention this involves.

Attention. As a child I learned to find four-leaf clovers as I walked along. A four-leaf clover is a break in pattern, a slight dissonance, that can only be seen against an awareness of the orderly configuration in the grass. The same kind of structure attention must have been what allowed my father so often to see a circling hawk or praying mantis or a moth, motionless on the bark of a tree, and point them out. You learn to watch for both harmony and dissonance. In ethnography, you also watch the people around you to see what they regard as ordinary and what they regard as unusual, and then you review your own responses because you bring your own biases and expectations. Then, if you are doing ethnography or natural history, you record carefully what your attention has allowed you to see, knowing that you will not see everything and that others will see differently, but recording whatever you can so it will be part of the cumulative picture.

What you look for in approaching a new culture is not so concrete as a beetle or a four-leaf clover, but the same search for order and dissonance goes on. At the simplest level, you are seeking regularity in behavior, but because you are an outsider you know none of the rules and expectations that others use to provide a framework of mutual intelligibility and predictability. To live in a village you need to know hundreds of details about who is who, who lives where, when people eat and when they sleep, and even these questions may be elusive. Gregory in his first fieldwork among the Baining of New Britain, before he went to the Sepik, found that a system of rules about who calls whom by which name—and which names cannot be pronounced at all—

made it almost impossible to sort out the relationships. At the same time, not knowing what to expect, he kept missing important ceremonies, sometimes deliberately tricked and sometimes caught unawares.

Fortunately, the details that seem so chaotic at first do hang together. It was Ruth who taught anthropologists to think about the almost aesthetic congruence of cultural detail, one step beyond the question of how it fits together in meeting the needs of survival and adaptation. Once you have a sense of the cultural style, the search for congruence and parallelism makes it possible to make cross-connections between one area of behavior and another. You notice, for instance, that the Balinese snack constantly and publicly— but mainly on liquid foods—while solid foods are surrounded with anxious preparation and taken furtively, facing the wall—and you connect this in your mind with the contrast between complete casualness about urination and secretiveness about defecation. You notice how play between mother and child is suddenly interrupted, the mother staring vacantly into space, and compare that vacancy with the way dancers go suddenly into trance. You find that the white cloth the witch uses in her dance is the one used to carry an infant, and start searching for other parallelisms between mother and witch.

Often a dissonance, the interruption in one pattern you have learned to expect, is the key to a larger pattern, as the routines of everyday are interrupted by a calendrical feast, but culture also provides the system of meaning for responding to the unexpected. There are no cultural rules that predict a psychosis, but it turns out that both the insane individual and those who have to respond to him and control him are working with cultural materials. In anthropology one is engaged generally in discovering the normal and the ordinary of another society, but whatever seems to be a dissonance in the context of the emerging order may provide a clue— including those things that are frightening or uncomfortable.

Margaret always emphasized the importance of recording first impressions and saving those first few pages of notes instead of discarding them in the scorn of later sophistication, for the informed eye has its own blindness as it begins to take for granted things that were initially bizarre. When something occurs to you, *write it down*, she said. Keep a carbon of every letter and every note, and send the copies home on different ships. *Date* every scrap of paper on which you write. Hoard the record of your queries and your errors, for the pursuit of any one may lead to knowledge.

Margaret and Gregory took anthropology a great distance forward in the recording of ethnographic detail, devising completely new ways of using film and still photography. They were very much concerned about documentation and about how to make fieldwork a more scientific process. In his early fieldwork, Gregory had gone through long periods of perplexity about what he was recording and why, and Margaret had already been criticized on her earlier books by those who questioned how she had been able to process so much material and how she was able to fit it together so neatly. She was one of the first anthropologists to prepare students for the field in a practical way, teaching them actual techniques.

Ideas that came into anthropology from psychoanalysis in the thirties strengthened the search for congruence. The theory with which Margaret and Gregory set out for Bali— about the relationship between infant experience and adult character, and the expression of character in the arts and ritual—led them to look for particular kinds of recurrent pattern. Very rapidly they felt they had found the key to Balinese culture in the handling of points of emotional intensity, where arousal is followed by frustration, and set out to document this, which they did more lavishly and elegantly than had ever been done before, in thousands of photographs. This sequence, from a sense of insight to a massive effort at documentation, is one in which subsequent attention has been shaped by a moment of recognition. It is justified by the conviction finally carried by the evidence collected, but another fieldworker might focus attention at some other point and come up with a different emphasis.

Margaret struggled for that moment of insight and felt that the limitations in the method would be transcended by better and better methods of observation and recording, so that repeated analyses would allow alternative interpretations. Gregory was dissatisfied when the insight was specific to a single culture and strained for a formal framework of description that would be deductive, onto which specific insights would be mapped. Indeed, in later years when Margaret was involved in simultaneous studies of a number of European cultures, he referred to her approach as "culture cracking."

By the time my conversations with Gregory had moved beyond childhood natural history and closer to his own intellectual preoccupations, he felt he had acquired the intellectual tools that would allow a true science of human behavior, especially the Theory of Logical Types and cybernetics. Onto these, he was able to map the key ideas he had developed in his Iatmul and Balinese research and in his work early in the war. The new subject matter to which he would turn, after the end of his analysis with Elisabeth Hellersberg, would be psychotherapy, starting with a collaboration with Jurgen Ruesch that produced the book *Communication: The Social Matrix of Psychiatry*, and going on into theorizing on schizophrenia as a disorder created by distortions in communication in the family. During these years, Gregory's particular intellectual style and bias allowed him to lead a research group right up to what has been regarded as a major insight into both the etiology and the treatment of schizophrenia, and then alienated him from that research as he rejected the changing structure of attention that went with institutionalization: increasing emphasis on therapeutic results and on empirical rather than logical demonstration.

Gregory had set out to introduce me to the Logical Types quite early—I associate the theory with a picnic in Central Park, probably when he was in New York seeking research funding. He got out a pen, and pointed out that this was something one could write with or put into a pocket. Then he proposed the class of pens, pointing out that the class of pens is not a pen: You cannot write with it or put it into a pocket. Then he proposed the class of non-pens, which clearly includes things like pencils and sticks . . . and automobiles and people . . . and the class of non-pens itself. Eventually, you can construct two new classes: the class of all classes that include themselves (like the class of non-pens) and the class of all classes that do not include themselves (like the class of pens). This seems harmless until you propose a class of classes, and then suddenly you are mired in contradictions, which are only resolved by systematically distinguishing the logical levels, so that it is recognized that the word *class* does not have the same meaning in "class of classes" that it had in "class of pens." If social scientists would keep the levels straight, they would not use phrases like "society forces the individual" or "history teaches," since neither is any more possible than it is for the class of pens to write.

When messages are at different levels, one message may operate on another, just as one class may contain another, with relationships more complex than double negatives. Gregory argued that this occurs constantly in mammalian communication. The message "I am about to attack you" may be framed by another message indicating "this is play," or the message "I am your mother," spoken to a child, modified by a message of posture or tone conveying rejection. If two messages conveyed at the same level of communication are contradictory, it is relatively easy to hold both in abeyance, to note the contradiction or to reject one side—but if the two messages are at different levels, one operating on the other, then a contradiction can create an oscillation like the paradox of Epimenides the Cretan who said, "All Cretans are liars." In saying this the philosopher declared that anything he might say was untruth—but what of the original statement? An oscillation is created unless you regard the original statement as different in kind, creating a sort of frame around all subsequent messages. Among us mammals, at least, Gregory asserted that there is always a frame which conveys a metamessage of some sort, a message about the message. There is no possibility of communication that is without context and logically flat.

Schizophrenics are often unable to sort communications in terms of metamessages specifying the difference between play and threat or report and promise, or to distinguish the metaphorical from the literal. Gregory would bring to his work with schizophrenic patients the assumption that their communication was not nonsensical or disorderly, but rather that it had an error in logical structure. It is critical, however, that the logical structure of an interaction is not limited to one person or to one moment. Gregory used to talk about the fact that if an organism is limited to analogical communication—to, for instance, acting out a message—then it is impossible to send a negative: You can convey a message that says, "I am about to attack," but not a message that says, "I am *not* about to attack"; instead, you must send the message of threat and negate it, perhaps by a metamessage that says, "this is play." The context of a communication, the frame, involves the therapist or the patient's family—over time, both the therapist and the family—in layers and layers of potentially contradictory material.

When Gregory first encountered the Theory of Logical Types after the war, he was gaining the tool that underlay almost all his subsequent work and yet it seems to me that he had already prefigured it in his discussions of learning, when he worked toward a distinction between "proto-learning" and "deutero-learning," learning to learn. While experimental animals perform the tasks set for them by psychologists at the "proto" level, they also learn something about experimenters and about varieties of learning experiments and may increase their efficiency because they have learned to learn. Here we are immediately in the presence of levels of abstraction, and can go on to think about even higher levels.

When I used to wander upstairs from my basement room, long after bedtime, complaining of "bad thoughts" that kept me awake, the grown-ups would advise me to go back to bed and think of something pleasant like a beautiful garden. "But I can think of dangerous monsters and rose gardens at the same time." "No, you can't—you can't think two thoughts at the same time." "Well," I said triumphantly, using an argument that Margaret and Gregory found inex-

plicably amusing, "I can *think about* thinking about monsters and roses; that's just *one* thought and it still keeps me
awake!"

It was the theory of deutero-learning that Gregory used
to account for the development of pervasive pattern in
behavior, since this is the process whereby premises about the
nature of experience are established—how others are moti-
vated, what the world is like, what degree of freedom or con-
straint characterizes it. Single contradictory experiences do
not destroy deutero-learned premises.

In 1964, when Gregory had moved to St. Thomas with
his wife, Lois, to join John Lilly's group studying dolphins,
Barkev and I went to the Virgin Islands to visit them for an
Easter vacation. I was at the time involved in an intense
exploration of the Episcopal Church, a return trip into Christianity that had begun in graduate school, and my father
mumbled and disparaged, saying that I should be ashamed of
myself, coming from a proud tradition of three generations
of atheists. We argued back and forth about how to count his
grandfather, William Henry Bateson, a clergyman who had
had a major role in changing the rule that all dons at Cambridge must be ordained. The argument went on at length,
one of those Bateson breakfast-table conversations that ate
away much of the day. Look, he said, the trouble with Christianity is that you cannot say to someone, you *ought* to love;
that's a third-order logical type, which means that it's not
intelligible, it's just nonsense.

I set out to create a demonstration that Gregory developed and used in subsequent years: If you say the Japanese
respect males, that's already at least a second-order statement, right? A generalization. And if you say they respect
males *more than* females that takes it one step further? And

then you say that, although the French also respect males more than females, the differences between the degree to which Japanese differ in their respect for males and females is greater than the degree to which the French differ in their respect for males and females? Right? Perfectly intelligible. And then you could go on and increase your sample of cultures or your types of disparity and compare disparities between disparities between disparities ... "WHOA, there, Cap, you're using words which summarize multiple levels, like disparity. You can do that with abstractions, but not when you are communicating about emotion."

A full ten years later, Gregory was recycling an argument that now seems to me to have a form very similar to his argument about love. Punishment, he said, cannot be used to deter or extinguish criminality. Punishment addresses crimes and can deter a specific action, but not "criminality" which flows from a learned general premise about the nature of action. If you put a rat into a test setting whereby its exploration will lead it to an electric shock, it will avoid the place where the shock occurred, but it will not give up exploration—after all, did not the discovery of the shock in fact underline the value of exploration? And similarly, the criminal who is punished may give up a specific behavior but still be encouraged to devise some other crime that he can get away with. An injunction to specific acts of virtue is very different from an injunction to love.

Gregory had a collection of favorite antibehaviorist or anti-rat-runner stories that he used to tell, most of which emphasized the failure of experimental psychologists to look at context, indeed to look beyond learning at the very lowest logical level. Another tells of a rat-runner, more biologically sophisticated than most, who reflected that rats do not in nature live in mazes, and so maze-learning experiments are not testing genuinely adaptive tasks. Instead, the thoughtful rat-runner procured a ferret, for ferrets in nature hunt in rabbit warrens which are much like mazes. He put a fresh haunch of rabbit in the reward chamber and released the ferret. The ferret went through the maze systematically, exploring each blind alley until he reached the reward chamber and ate the haunch of rabbit. The next day, the experimenter started him off again to see how quickly he would learn his way. Again he went systematically down every alley until he came to the one that led to the reward chamber; but he didn't go down that one—he'd eaten that rabbit. What he learned in the framework of the learning experiment was set in the context of his premises about how the world works. I asked Gregory where the experiment had been published and he said as far as he knew it hadn't been—the experiment was regarded as a failure.

Gregory spent several years in the mid-sixties studying dolphin communication in the Virgin Islands and then later in Hawaii. At one point I was offered the chance to go into a pool with Peter, a young male dolphin.[8] Dolphins are known to be solicitous of human beings, rescuing them from drowning and not attacking them even when attacked by them, and Gregory argued that just as many mammals (including, to a considerable degree, human beings) are inhibited from attacking the young of their own species, the inhibiting signals for dolphins that protect their own young must also apply to human beings—or, to put it differently, dolphins are reminded by people of cubs and treat them accordingly.

Treat them, indeed, like notably stupid cubs. I went into the pool with a snorkel and my moderate swimming skills, waiting quietly for the dolphin to make his own advances.

More than six feet long, with a huge grinning mouth open wide, he charged across the pool, only to brake and turn with amazing control, racing away and back, and then whirled around me so I was caught in a current of his creating like a toy in the water, and spun, and then gradually he would make more frequent contact, like a cat that rubs against one's ankles, soft and luxuriant, dolphin skin shivering at human touch. The condition of going into the pool was the keeping of a record, so Gregory was there with his camera and Lois with a notebook, and after I came out I went home to write an introspective account of the extraordinary experience of being reduced to helpless infancy and simultaneously courted by this seductive creature of another species, establishing trust by the repeated analogic sequence of threat—the gaping jaw and the charge—and the denial of that threat.

All of this I tried to get down on paper the next day, for it seemed to me that I would be more able to put my own mixed feelings on paper in the freshness of that first experience than others who approached with a scientific identity to defend. "How on earth," I asked Gregory, "can anyone remain objective working with dolphins? How do you keep from being sucked in?"

"The same way I do with Iatmul or Balinese or schizophrenics," he said. "You can't work with human beings without allowing for your own involvement. But biologists don't know that—they're used to working with fish or birds, all creatures that don't try to seduce them—they are not able to observe themselves in the relationship, so they produce nonsense."

Later in Hawaii he told a story of a learning experiment, playing catch with a hoop, conducted with one of the dolphins where the experimental protocol involved a number of repetitions that any experienced dolphin trainer would have warned would be boring, so the animal would break the sequence by tossing the hoop off to the side. "Funny," said the assistant to Gregory, "every time he does that he gives a little chuckle." Gregory asked if she wrote that down, but there was no place to record it—only that at that point the dolphin had failed to demonstrate successful learning.

The problems of attention and of disciplined subjectivity are part of the problem of consciousness—knowing and knowing that you know. If you can draw a frame around an event, you can briefly separate it from context, and this is what many researchers believe they are doing. But an anthropologist in the field cannot generally do this, and must assume that the asking of any question shapes the answer. Even as you observe, you also participate. The context of any question, the entire conjunction of interviewer and informant, sets a metamessage for the communication. Trying to be objective, you may think you are separating off an experience by setting it in a frame, but actually the frame changes the meaning of what is within it.

It is not difficult to look back over work done by Margaret and see that sometimes it would have benefited from another layer of self-consciousness and self-criticism on her part, but in general it was her very engagement that made insight possible. Over time, she was very explicit about methodology but less given than Gregory to puzzling about the structure of thought. Nothing in her work corresponds to *Naven*, Gregory's book on the Iatmul, in which a very small quantity of data provided the occasion for a series of reflections on theory making. His ruminations in *Naven* about the complex organization of culture are still challenging students of anthropology, for culture is a human product and the mind of the scientist is itself a human mind. Every culture develops over the generations in such a way that an out-sider—the child born into it—can come to understand it.

One of the premises of the household in which I grew up was that there is no clear line between objectivity and subjec-tivity, that observation does not preclude involvement. The floodlights that stood ready to make photography possible were tokens of that concept. There was no rule that said that sincerity is expressed by inarticulateness or that to be analyt-ical about an emotion or a relationship in some sense violates it. A close friend once complained to me that Gregory was too exclusively analytical and unconcerned with feeling, but in fact his conviction that emotions are patterned and logical meant not that he was exclusively rational but that he gen-uinely rejected the notion of a separation between thought and feeling. "A tear is an intellectual thing," he quoted, and David Lipset used this as the title for the first version of his biography of Gregory. Gregory was fascinated by the clever-ness, the logic and elegance, of emotion and of relationship, frequently quoting Pascal: *"Le coeur a ses raisons que la rai-son ne connaît point."*

When I was pregnant, I decided to use the altered atten-tion that would accompany new motherhood by involving myself in a research project on mother-child communication. Between breast-feedings I spent hours in a tiny projection room at MIT's Research Laboratory of Electronics studying films and tapes of mothers and newborn infants.[9] Sometimes, especially in the beginning, time would pass and the accus-tomed hour of a feeding approach, and baby noises on the tapes would hasten my readiness to nurse as milk stains spread on my blouse. The moments of play I began to enjoy with Vanni at home made me able to pick out a series of very brief passages in the films that could then be analyzed frame by frame to demonstrate the establishment, within a scant month after birth, of the kind of alternating interaction I later called "proto-conversations." I also did acoustic analyses of the infant vocalizations, of a type not then described in the scientific literature—too slight, no doubt, too insignificant to have attracted the attention of acousticians, but significant to me in the heightened attention of new motherhood.

My own experience has always been grist for the mill of analysis, and analysis has always been for me a way of dealing with emotion. I wove the way Filipino friends reacted to the death of our premature son, who was born and died in Manila, into an analysis of bicultural contact,[10] and I dealt with the stress that surrounded my departure from Iran during the revolution through an interest in the way Iranians moved between personalizing and depersonalizing such moments. There is indeed a sense in which this book is the latest of many phases of trying to relate personal experience to abstract ideas, assuming that knowledge and art and caring are all intertwined.

ANTHROPOLOGISTS FROM NEW GUINEA.

From left: Mr. G. Bateson, Dr. Margaret Mead, and Dr. Reo
Fortune, who arrived from New Guinea yesterday by the Macdhui.

Sydney, Australia

Ruth Benedict

Courtesy Vassar College

Mosquito room in Iatmul

G. Bateson

The Museum model of old Pere Village

Classifying Balinese carvings

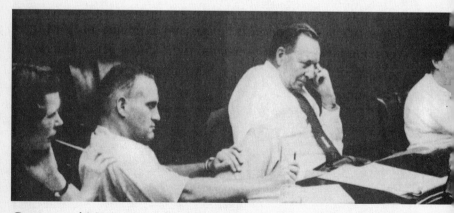

Gregory and Margaret participating in a conference that illustrated a book on conferences

Karl Heider

Paul Byers

A village council in Pere

Barbara Roll

Interviewing J.K.

Barbara Roll

Cecil Beaton

PORTRAIT OF MARGARET
REACHING OUT TO THE WORLD

TRANSIENT MEETINGS WITH GREGORY

Discussing *Our Own Metaphor*

Polyxane Cobb

Left: Gregory
with Lois;
Below: Gregory
with his three
children, John,
Catherine, and
Nora, Barkev
standing
alongside

Lois Bateson

PORTRAIT OF GREGORY AS ONLOOKER

ACROSS THE GENERATIONS

Polyxane Cobb

Catherine, Barkev, and Vanni with
Gregory, Margaret, and Aunt Marie

Ken Heyman

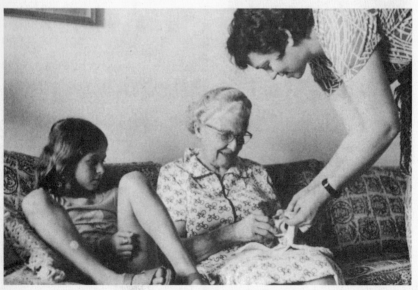

Beryl Bernay

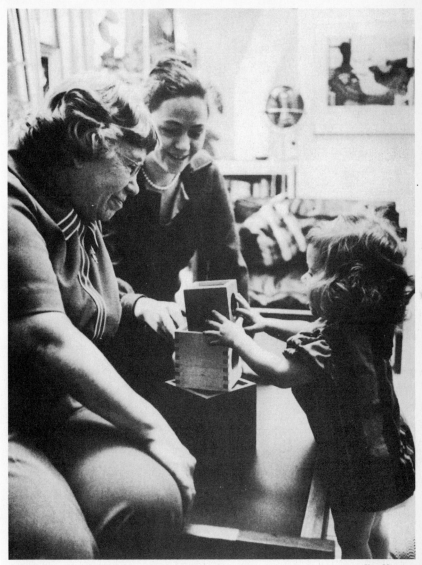

Ken Heyman

Fred Roll

A LAST ENCOUNTER

XII
Our Own Metaphor

My first real intellectual collaboration with Gregory came at the point where he had decided to try to take a hand in shaping events instead of sitting back and watching the world "going to hell in a handbasket." He had agreed in 1967 to plan and lead a conference to explore the way in which human habits of thought may be leading toward planetary disaster. In doing this he was abandoning a stance he had maintained since the war when his discouragement about his own contribution and his distaste for applied social science had been a factor in his depression, but I think his reluctance to attempt to influence events went far deeper than that particular sequence of experiences in expressing a pervasive fatalism, for he had all his life the belief that efforts to solve problems are likely to make them worse.

It may be that at the point where he began to want to think about affecting the course of events he turned toward me as possibly able to provide counterpoint, as Margaret had done in their joint work. She and he contrasted very sharply on exactly this dimension of personality, the assumptions they made about what could and could not be changed or improved. Margaret believed that problems can be solved. Not only had she learned particular actions to alleviate particular problems one by one, she had also learned to assume that a solution exists and had developed strategies for seeking it. "We need a social invention," she would say. The same general premise pervaded her approach to global issues like the question of nuclear disarmament and family questions like my concern that I could not do as much Armenian cooking as Barkev would have liked. Her strategy was to search for a point of leverage, often a partially symbolic one, or a new way of phrasing the problem that might suggest solutions. To me, she suggested that I keep a soup pot going, which would fill the house with a good smell of cooking and a sound of bubbling, symbolizing domesticity to my husband. Thinking about the nuclear danger, she invented the phraseology that only we can protect the children of the Soviet Union and only they can protect our children.

Margaret's approach must have been based on early success in dealing with problems, perhaps related to the experience of being an older child and amplified by years of successfully organizing the younger ones. Gregory's experience was that of a younger child with relatively little capacity for changing what went on around him. Instead, he would seek understanding. Indeed, he had a kind of abhorrence for the effort to solve problems, whether they were medical or political. It may be that the suicide of his brother Martin in 1922, which followed on heavy-handed parental attempts at guidance and led to a period of increasing efforts to shape Gregory's choices as well, was an ingredient in his anxiety about problem solving and indeed about any effort to act in the world: He was as dubious about active intervention in therapy as about politics or economics.

Basic premises of this sort are self-ratifying—one's tendency is to interpret any sequence of events in such a way that the outcome confirms one's expectations. Gregory came out of World War II feeling that he had made no significant contribution and convinced that applied social science was necessarily dangerous, while Margaret came out convinced of the value of her own work and of what could be achieved by putting social science to work for human welfare.

It is not difficult to find instances for each where these particular patterns of assumptions were inappropriate. Margaret tended to approach her own body as she approached other problems, taking a substantial range of medications, dosing herself to maintain her performance at maximum efficiency. When she was dying she was continually trying to find the point of leverage that would help, either in reversing the disease or in alleviating her discomfort. Gregory was routinely skeptical of the effects of medical treatment and neglectful of his body. He let his teeth rot and drop out, unreplaced, and a festering wound in his foot went untreated for almost a year. He told of how a dentist had advised so much work before a field trip that he decided none of it was worth doing, and how most of it was apparently no longer necessary on his return. Dying, he struggled to understand how to relax and let go.

On the other hand, each of them often found ways to compensate. Margaret knew that her tendency to try to find solutions could trigger resentment because it was experienced as manipulative, while Gregory knew that he needed colleagues who would counter his tendency to be fatalistic, so he mixed a disdain for the entrepreneur with a reluctant admiration for those who could get things done. One can see how this kind of complementarity made it possible for Margaret and Gregory to work together productively and insightfully—and how infuriating and exasperating each could be to the other.

The change in Gregory's stance and his interest in planning a conference came about as a result of a number of different processes. A new consciousness of the dangers of environmental pollution was spreading through society in the late sixties which led to renewed awareness of the danger of nuclear disaster that has been with us increasingly since 1945. The two kinds of danger have formal similarities which Gregory had characterized as early as 1946 in terms of regenerative feedback, describing systems that go into runaway instead of having built-in feedback loops that give them the capacity for self-correction.

The difference was that now there was beginning to be an audience for what Gregory had been nursing to himself for a very long time. Thus one of the things that changed and helped Gregory to become involved was the nascence of the counterculture. The political engagement of the early sixties was not appealing to him, but the culture of disengagement was. Particularly on the West Coast, Gregory was becoming known and admired as someone who might provide alternative intellectual approaches. Alienated, he was not uncommitted, for his rejection of the surface phenomena of social life was balanced by an allegiance to underlying patterns that he loved as much for their elegance as for their role in sustaining viability.

The beginning point was to be a conference in the summer of 1968, in Burg Wartenstein, a castle in Gloggnitz, Austria, owned by the Wenner-Gren Foundation for Anthropological

Research, where every summer a series of conferences was held following the foundation's distinctive model. The conferences were small, limited to twenty people, and sustained, with each participant present for the whole period of up to a week. Gregory put together a group that included individuals with backgrounds in very different disciplines—psychology and intellectual history, mathematics and psychiatry, neurology and cybernetics and ecology, anthropology and linguistics—and when on the first evening we reviewed the intellectual histories of the group we found that almost everyone there had at some time shifted like Gregory from one field to another. Papers were prepared and circulated in advance so they would not have to be formally read. Every aspect of those days was designed to facilitate productive talk, the kind of talk in which the participants not only learn new things but are occasionally able to say something which no one of them knew as they came in through the door. My father wrote to me in the Philippines, where I was living, and asked me to serve as rapporteur, referring to the fact that I had, as a graduate student, edited the proceedings of the formative conference on semiotics.

I think back over the sequence of conversations between us that tracked Gregory's transition from abstract communications theory through the study of schizophrenia and then to the study of dolphin behavior, and I realize that even before the conference in Gloggnitz Gregory and I already shared an idiom, sustained and developed from the natural-history lessons in my childhood, that allowed me to slip with him into a form of argument that was also a dance. We had very little time together between 1960 and 1968, but a thread of conversation continued—biology, anthropology, linguistics. Once on an adult camping trip he asked about the current state of thinking on what is called the Sapir-Whorf hypothesis, the hypothesis that there is a causal link between thought and language, so that the patterns of thought of speakers of different languages differ. "I suppose," he said, "that it's one of those things that cannot *not* be true." I agreed but pointed out that efforts to prove it were unsatisfying. "Get it said right," he said, "and then it will be self-evident." There are certain truths that cannot be said so as to be understood without being believed, and one of his concerns was to formulate the nature of human conflict with natural patterns so that the need for a change in direction would be self-evident.

My report of Gregory's Conference on Conscious Purpose and Human Adaptation was eventually published under the title *Our Own Metaphor*. I was trying in that book to lead the reader into a number of complex and unfamiliar ideas, and one of the ways I did this was by exaggerating my own difficulty in following the conversation, easing the reader into discussions of regenerative feedback or entropy through my own musings, presenting them as they had first been explained to me as a child—entropy, my father had defined as "the tendency of everything to get into a muddle." I introduced the book with a description of building, with Gregory, my first aquarium. Just as he did, I started with a description of the thermostat that we bought to switch the heater on and off and keep the temperature within the appropriate range as an example of cybernetic governance by constant correction, and went on to talk about how we tried to simulate the natural balance of different plant and animal species in interaction. In retrospect, having portrayed myself so often in that book as grasping this or that concept for the first time, step by step, I find it very difficult to recover a sense of what I did know. It seems to me that I arrived in Burg Wartenstein less diffident than I later painted myself, but I know that I was deeply pleased that Gregory had read and admired my paper. "Now, this makes sense," he said. "The girl can think."

Already when I arrived, I had concluded that I did not want to do anything like what I had done in reporting on the semiotics conference. I wanted to write a book that would read like a novel, so that the reader could share the experience of participation. I had, after all, spent my life in and out and on the edge of conferences, formal and informal, treating them as a normal mode of interaction—perhaps indeed as *the* normal mode of interaction.

Barkev and I, in the early sixties, had become involved in a series of conversations about a daydream which we called the Center for Correlated Studies or, for short, "the Egg," a daydream triggered by the notion of a wealthy friend of ours of acquiring a particularly lovely old mansion on the Massachusetts North Shore and turning it into a conference center. Barkev had been interested in some of the early work on group dynamics, while I had grown up to believe that conferences are the way to think. We designed a structure that would function on two levels, gathering a group of people to work together on particular themes, and always at the same time having a discussion of the process, a process in which interpersonal relationships as well as ideas are interwoven. We believed that these two levels could interact and that instead of creating an uncomfortable degree of self-consciousness, what we would be creating would be a liberating form of insight.

Curiously, the Gloggnitz conference was echoed by a conference convened at the same time in Alpbach, Switzerland, by Arthur Koestler to discuss many of the same issues. Koestler believed, however, that the resolution to the problems of human destructiveness would occur by the creation of new physical or chemical connections within the brain, which Gregory, concerned with the same problems, ultimately felt was mechanistic. Each invited the other; neither could attend because the dates conflicted. After Gloggnitz, Gregory and I spent a week touring around Austria and Switzerland, continuing the conversation, and we went to visit Koestler in Alpbach, comparing conferences and comparing what we intended to do.

There were eventually two books based on the Alpbach conference, for in addition to a conventional conference proceedings, Koestler wrote a novel, *The Call Girls* (which incorporated some of the anecdotes we told on his veranda that evening about our own gathering). We were elated and he was ambivalent; *The Call Girls* ultimately expressed his anger and disdain, if not for the real participants in his actual conference, at least for that kind of person and for the process itself. Koestler set up two different processes, one called truth and one called fiction; I reached the conclusion that my book would be true to the event only if it followed some of the conventions of fiction. Most important, I would not write an "objective" account of the conference, because by the conventions of academic reporting this would mean editing out emotions that seemed to me essential to the process. The emotion was edited out of the formal proceedings of the Alpbach Symposium, which came out dry and academic, and resurfaced in the novel as rage.[1]

We had a good deal of rage around our conference table in Gloggnitz, as well as moments of love and of bewilderment. There is a sense in which the emotion was edited into the *Metaphor* book, for I used my own introspective responses of dismay or illumination to bring the reader into the room, and worked with the tape-recorded discussions so that the emotionally pivotal comments would be brought out rather than buried in verbiage. The book only reproduced those portions of the scholarly presentations of the participants that became relevant to the emerging directions of the thought of the group and to the interpersonal event. One member, Tolly Holt, has said that he thought that Gregory "tried to create spokesmen for the various aspects of his own psyche,"[2] but some members of the group felt they were miscast and others objected passionately to the way in which what they said was responded to by others, twisted to fit into a plot that was not fully shared.

Burg Wartenstein is a fortress built high in the mountains guarding the Semmering Pass, with great gates to seal off the inner keep and a rough flagged courtyard within the shelter of the walls. A conference such as ours becomes a world in itself, sealed off in its own self-definition, as the participants are lifted out of their normal lives and backgrounds and forced into the effort of mutual adaptation. One is held in an envelope of time and inaccessibility, like the glass sides of an aquarium, as different kinds of minds work sometimes toward conflict and sometimes fall into a sort of dance of symmetry or counterpoint that leads to moments of revelation.

The beauty and natural peace of the environment were a background against which most of us realized for the first time that it is truly likely that in one way or another human actions will destroy life on this planet, and this gave us an aching sense of urgency. Strangely, this fairytale castle, from whose towers one could look down into the manicured woods and farmlands of the valley, was for us not so much the ivory tower of anti-intellectual mockery but a place to dream of the role of guardian, a watchtower of protection. There were moments when we believed that by the way we ordered our discussion we could actually move toward some resolution of problems; indeed, some members of the group were ready to try to plan action campaigns, while Gregory stubbornly insisted on the priority of the search for understanding.

Margaret used to say that a really successful conference is one in which the intensity is so great that you feel as if you are falling in love. This intensity was compounded at Burg Wartenstein by the fact that at least three generations of intellectual relationship were represented there. Warren McCulloch, lean and gnomic in all that he said, belonged to the grandparent generation, one of the elders of cybernetics from whom Gregory felt he had learned most, his craggy bearded face already prefiguring his death. I was the youngest present, but there were several young protégés of Gregory's or Margaret's. Ted Schwartz was there, who had accompanied Margaret to Manus in 1953 as a graduate student, selected by her both for his abstract turn of mind and his willingness to engage with photography and other kinds of equipment, almost as a replacement for the kind of counterpoint to her own thought that Gregory had provided before the war. And there was Tolly Holt, an intense and idealistic mathematician whom Gregory and Margaret had listened to when he was a teen-ager and said, "In ten years we won't be able to understand what he's saying." Tolly played the role of son, resisting the effort to be dispassionate about the fate of the earth.

Tolly and Ted, each in a different way, challenged Gregory like the young males of the pack circling around the leader and displaying their increasing strength, so that intellectual disagreement acquired all the passion of familial drama. It is not surprising that after the intensity of those days, when we traveled off together, Gregory fantasized a romance between himself and me. I looked at him and thought, no, the whole structure of the drama moved me in love toward Tolly; in stories, it is the prince who gets the princess. Yet all of this sharpened our intellectual involvement.

The conference was called to discuss a single ominous proposition that Gregory had formulated. As Tolly said, it had "the form of a scientific question but the substance of a warning cry."[3] *"That the cybernetic nature of self and the world tends to be imperceptible to consciousness, insofar as the contents of the 'screen' of consciousness are determined by considerations of purpose."*[4] To the extent that this proposition is true, the destructive—perhaps fatal—impact of modern technology on the environment will continue.

It took us the first several days even to understand what Gregory had in mind. Over the years Gregory often started his efforts to introduce others to his thought by trying to make them aware that perceptions cannot be seen as identical with physical phenomena, and that one cannot be aware of the transformations that lie behind one's own mental pictures—the pervasive problems of epistemology. I remember being horrified as a teen-ager when he pointed out, holding up a pack of Luckies (which he smoked between the pipe and the emphysema), that there is no way to know that the experience I have when I see the color red is the same as the experience that another person has—there is no way to put those experiences side by side. What the senses report to the screen of consciousness is not a physical object existing in the physical world, but a series of differences, from which a mental map of experience is constructed—but the map is not the territory.

Gregory was proposing a particular and lethal structure to the distortions of perception: that consciousness, shaped by purpose, distorts perception in a specific way, making us think that the world works in lineal sequences. We believe that we can go from *A* to *B* to *C* in achieving our purposes, but in fact each step has multiple effects.

The conference proposal was largely a statement of what Gregory had come to mean by the "cybernetic nature of self and the world," his statement after thirty years of mulling it over of the paradigm he found at the cybernetic conferences of the Macy Foundation. He had learned there a way of thinking about individual organisms, societies, and ecosystems as complex assemblages of interrelated parts that depend on internal feedback loops of communication to maintain certain truths about themselves, as the body maintains its temperature and as the populations of predators and prey are maintained in balance in the forest, in spite of fluctuations. But the insight into the nature of stability and balance that cybernetics brings, namely that each is achieved by a constant process of correction across a multitude of variations and oscillations, the poise of the tightrope walker, is only the first step. This process of adjustment to achieve homeostasis takes place in a complex system at many points, conserving the values of many different variables, in multiple interconnected loops, and within any such system there are points of vulnerability and subsystems with the potential for runaway. Causation moves in circles rather than in simple straight-line sequences.

An example of what Gregory had in mind as the lineal logic of conscious purpose, which is familiar to most people today, is given by the early uses of DDT. In order to increase harvests, farmers tried to control insect pests by spraying:

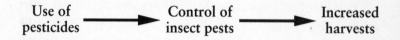

The reality was not a straight line of causation but a circular system in which it turned out that what the use of pesticides actually does over time is to increase the use of pesticides. Under spraying, insects are subject to an accelerated natural selection that produces resistant strains, while their natural predators, including song birds, die off. Natural selection works as a self-corrective process that maintains the insect population. Because insects are born and die in vastly greater numbers than song birds, the process of self-correction is more rapid at their level of the system.

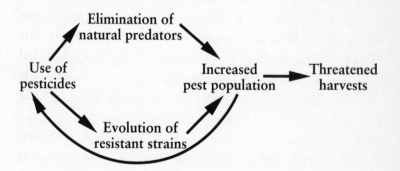

This is of course only part of the system, for it leaves out another self-reinforcing circuit in which the use of pesticides stimulates the pesticide industry—which also stimulates the use of pesticides. It also omits other unanticipated side effects—some pesticides, for instance, turn out to be carcinogens. The farmer and the pesticide manufacturer, in pursuit of their conscious purposes, have seen a complex interactive system in terms of a single line of causation.

Gregory came to see many kinds of progressive change or breakdown as expressions of the systemic effort at constancy. Small alterations can ramify through the system in compensation for local perturbations, so that gradually changes occur that modify the nature of the system; occasionally parts of the system go into runaway and the whole breaks down. If you think in lineal sequences, the unintended results are dismissed as "side effects," but they return to resonate in the whole. In the years following Burg Wartenstein, much of Gregory's eloquence came from the use of this single model to compare different types of change: learning, addiction, evolution. The words vary depending on what kind of system one is looking at and on the level at which one looks, the logical and organizational levels with which Gregory was most at home: A cell is a system and at the same time, of course, a part of a larger system, an organism, and organisms are in turn parts of ecosystems.

One of the reasons systems tend to have the capacity to go into runaway is that they are generally "set" to maintain some variables at excessive levels: All species produce more young than are needed for replacement in a constant population, young that are weeded out later. The whole enterprise of modern industry grows out of the human calibration to strive for more than just enough to avoid hunger and cold, coupled with the process of habituation that says that any less than I now have is too little and is experienced as privation. Thus Gregory was able to argue with the Regents of the University of California, when Jerry Brown appointed him to that body, that the way in which income from nuclear research had become intrinsic to the health and functioning of the university was a form of addiction.

Cybernetics, it seems, makes poets of us, as does any formal system that allows the recognition of similarity within diversity. If you look beyond a tree's rigidity to see it as alive, then you see it as more like a woman than like a telephone pole, perhaps indeed, in a slowness of growth that only seems static, living the impassioned life of a dryad. It is the same kind of looking that recognizes the spirals of growth in shells as the frozen form of cyclones and galaxies. Gregory, with his multitudes of arguments based on logical levels, his metamessages and metacommunication, was delighted when the quip went around: "Hitch your wagon to a star or what's a metafor?" The pervasive play of his discourse was indeed the play of metaphor.

At Burg Wartenstein I felt that in some serious way we needed to link the human reliance on metaphors grounded in bodily form and bodily experience—metaphors of eating and defecating, making love and giving birth and dying—with the kind of metaphor that takes a formal analysis like cybernetics and allows us to see the same pattern in the life of a group or a cell or a coral reef. Certainly there is an isomorphism—a formal similarity—between the process of repression that Freudian theory posits, where certain forms of self-awareness are concealed from conscious access—precisely those that deal with the most basic bodily processes and desires—and the failure to see the linkages of the natural world. I argued at the end of the conference that we will live at cybernetic peace with our environment only when we can acknowledge our own death.

Margaret used to list for her classes some of the ways to

achieve insight, with fieldwork among a primitive people and psychoanalysis high on the list. She included the study of animals and infants as well, perhaps because animals and infants are like and unlike oneself in particularly disturbing ways, ways that demand new definitions and new ways of caring. Each such experience provides access to additional layers of metaphorical thinking. Other experiences she listed were the ones in which one enters and thinks within a total system, distinctively ordering the world, and then steps out of it—a psychosis, a religious conversion, a love affair with an Old Russian—perhaps even the engagement and withdrawal of participation in such a conference. Her list did not include the study of more disparate systems, like a lake, yet this too can be a path to insight whether for a Thoreau or a contemporary limnologist like Evelyn Hutchinson, who went collecting beetles with John and Martin Bateson when they left their little brother, Gregory, behind to go away to school. Passionate attention to a lake, with its respiration, its maturation and aging, its dependence on interlocked forms of balance, its reflection of the self, is equally puzzling, equally revealing, equally a beginning of love.

During the conference, some real progress was made toward a description of the kinds of error built into conscious thinking about cybernetic systems, some errors perhaps common to all human beings and others developed or accentuated in the history of Western thought. Gregory started from the tendency, especially in the pursuit of consciously defined purposes, to see only sections of the circuits of natural systems and to treat these as straight-line cause-and-effect sequences.

As the conversation progressed it became clear that this focus on parts of circuits was one aspect of a far more general cognitive strategy of Western culture in which, in the effort to deal with complexity, complex phenomena are separated or dissected. From this point of view there is a similarity between atomism, the effort to understand complex wholes by looking at the parts, and body-mind dualism which historically simplified the study of material bodies by setting aside something mysterious called "mind" as belonging to theologians and philosophers so the natural scientists could get on with their business. The difficulty with such a strategy, which has been immensely powerful in the physical sciences and seems at first to be highly effective, is that in the biological world the analog of cause is difference. To understand a living system, it is necessary to look at a constellation of factors, not in and of themselves, like single moving billiard balls, but in their relationships·and in their contexts.

The notion of individual actors as causes, seen apart from their relationships, is an example of the same problem. Western individualism breaks up circuits—the circuits of birth and death and of phylogenetic change, the circuits of ingestion and elimination that must fit into natural circuits of growth and·decay, and postulates an absolute value in the individual.

As the conference progressed, we spoke increasingly in terms of premises, false premises about the natural world and the nature of action. This, for Gregory, was the beginning of optimism, for it located the problem not in a necessary structure of consciousness but in learned assumptions and patterns of thought. The didactic quality of his last book, *Mind and Nature*, is a direct product of this shift in emphasis as his distaste for muddleheadedness and slovenly habits of thought became the focus of his sense of danger. "Cap, why can't we get them to think straight?" What parable, what formalism, what ritual would provide a new pattern? Increasingly he felt that poetry and emotion and ritual were the places to look because they are sometimes less falsified by inappropriate simplification.

In the end, Gregory's sense that the conference was a major breakthrough had to do with a double structure: Our awareness of conference process came to inform our awareness of the issues of the environment. Gregory resisted proposed modes of action, out of his conviction that the attempt to achieve conscious purposes is the source of the problem. But apparently he came to believe that the conference itself had come to embody a process that might supply a solution, by weaving together the conscious and unconscious thought of many minds in a single system. Here we were at that point where the process in the microcosm mirrors the larger process and where intimately shared experiences are a source of broader knowledge. We were in a sense playing out the concept that Gregory had used in writing the series of father-daughter conversations he called "metalogues": "A *metalogue*," he wrote, "is a conversation about some problematic subject. This conversation should be such that not only do the participants discuss the problem but the structure of the conversation as a whole is also relevant to the same subject."[5]

I spent more than a year working on *Our Own Metaphor* as a vehicle for conveying these ideas and making cybernetic ways of thinking more accessible. Two aspects of the book meant that it dated very rapidly. One was the use of topical material to talk about ecological danger, material that was shocking then but soon became trite. Today we may be moved to a comparable sense of urgency only by the specter of a nuclear winter. The other was the critique of DNA theory as an example of atomistic thinking.

It may well be that all that was said about the style of early DNA theory was in principle correct, but the argument was intended to apply to all of molecular genetics. Gregory's readiness at Burg Wartenstein to listen to an attack on DNA theory was a curious echo of his father's skepticism, a generation before, about chromosomes, as somehow too material an expression of genetic process. Gregory was perfectly comfortable in his turn with chromosomes—one of the earliest "teach me something" conversations I remember concerned amoeba reproduction, with diagrams on a paper napkin of the steps of mitosis, the chromosomes sketched in like hairpins—but he was skeptical of some of the early formulations of the role of nucleic acids.

He used to tell the story of the "flat-earth man," who, lecturing on the earth, explains that it is shaped like a disk and balanced on the back of a turtle. Where, he is asked, is the turtle? Oh, the turtle is standing on a rock. Where is the rock? Oh, it's balanced on another rock . . . and it's all rocks *from there on down*. Those who seem clearest today that there is no basic material level to which everything can be reduced are those who work within the heart of the atom. Logically, there was no reason for Gregory to reject arguments based on molecular biology except when these are posed as statements that a particular phenomenon is *nothing but* an expression of biochemistry. He was clear in his rejection of the supernatural, of any transcendent reality, but he was equally clear in his rejection of simple reductionist materialism. It was in organizational complexity, the relationships between material parts, that he found the matter for almost religious awe.

Gregory called the range of questions on which he

worked in the last decades of his life "an ecology of mind."
He was developing an argument that sited human experience
and action ever more intimately in the natural world, and yet
affirming at the same time a discontinuity between material
and mental patterns of causation. Mind is immanent in certain kinds of material complexity, indeed in any system of
parts that is able to recognize differences which themselves
have no physical existence, and its response is energized from
some source other than these differences, often by stored
energy. A thermostat is a mind—it can recognize a difference
between its setting and an actual temperature and respond to
that difference by switching a heating system on or off. A
man may be angered by a courtesy he does not receive—the
difference between what he expected and what occurred—
and when he retaliates with a blow, the energy for his action
comes not from the discourtesy, which has no physical existence, but from his breakfast. A mind consists of closed loops
or pathways along which messages of difference are transmitted; its self-adjustment may be either in the direction of
homeostasis or in the direction of runaway.

It was of central importance to Gregory to think of mental systems as hierarchically organized—as exemplifying the
logical structure of levels of abstraction that so pervaded his
thinking. From Gregory's point of view, it does not matter
whether the system in which mind is embodied is a machine
or a single living organism or a multitude of organisms, a forest or a family or a conference. With sufficient logical complexity a mental system may become able to learn. Thus, for
him, mind could be recognized in an ecosystem as well as in
an organism, in a group as well as in an individual. The
boundaries of individual organisms tend to mark off variations in the complexity of parts, so that within an envelope of

skin or membrane the news of differences is physically transmitted in different ways and the self-corrective loops are denser. Depending on the boundaries and the time scales, the processes he was talking about are called life or learning, culture or evolution.

It was for him an absurdity to argue about whether a particular behavior was physical or communicational, since no mental event is exclusively physical and yet it must occur within a physical system of the appropriate complexity. Thus, when his work on schizophrenia was criticized, he defended himself from the either-or of physical and mental explanation by saying, "The appearances of schizophrenia may be produced by parasitic invasion and/or by experience: by genes and/or by training. I will even concede that schizophrenia is *as much* a 'disease' of the 'brain,' as it is a 'disease' of the 'family,' if Dr. Stevens will concede that humor and religion, art and poetry are likewise 'diseases' of the brain or of the family or both."[6]

As always, Gregory's style moves toward abstraction and toward the highest level of generality, and is expressed in ways that contrast sharply with what Margaret had to say. Still, the thought of both is centered in an affirmation of learning and the sharing of learning as the distinctive human possibility, an affirmation of human beings as at once physically and communicationally complex, capable of purposes and choices and yet embedded in larger systems of which they are parts.

Gregory's ethical urgency emphasized complexity and the dangers of manipulation to maximize some characteristic or support some addiction, his examples ranging from the administration of drugs to schizophrenics to the use of pesticides to allow the farming of monocrops, manipulations involving dangerous oversimplification. Again and again during the conference, he broke up efforts to write prescriptions for action, arguing that the value of the discussion lay precisely in its diversity and complexity, and that the significance of what had been said lay in the fact that each topic had been discussed in the context of the others. He made it clear that preserving and communicating that weave was the essential next step, and that was the task he passed to me.

I went back to Cambridge in the fall with my head buzzing. My role as rapporteur meant that I spent months with the transcribed discussions from Burg Wartenstein and with the tapes themselves, hearing again the familiar voices talking about molecular biology or animal training or mathematical representation, about the Paliau movement in New Guinea and the movement to ban atmospheric nuclear testing, about computers and the population explosion and the impact of Descartes on philosophy—either a terribly muddled hodgepodge or a model of the range of materials we will have to be able to use with precision and with awe to understand the problems posed by the kind of animal we are, living on this kind of planet. Sometimes as I worked I became terribly discouraged, feeling that Gregory's arguments must lead inevitably to nonaction, and yet encouraged by his emerging commitment to communicating them. Given appropriate models, it is possible to act to increase or protect the diversity that gives natural systems their resilience, and it is possible to teach people to attend more deeply and more broadly, to advance in understanding and to watch for the unanticipated dangers of efforts at solutions.

XIII
Chère Collègue

If you consider the actions that Margaret felt able to advocate, from many of the same root convictions as Gregory, you see that what she characteristically did when she acted in living systems was to enrich and elaborate rather than to simplify—to offer additional metaphors, more communicational links, a richer level of awareness, and to introduce greater diversity. As Margaret moved around the world, engaging in conversation after conversation, she was a one-person conference. She carried a little notebook around in her purse at all times, writing down any new idea or information she thought she might want to use.

The notebook stands in my mind for a whole way of working whereby she was constantly taking in new material and using it, incorporating reactions, so that an interesting piece of work she heard about in Florida would be talked about in Topeka and synthesized with what someone was thinking or writing in Boston, elaborated in Cincinnati, incorporated in a lecture in California. She tried to be conscientious about giving credit and would often put people working on related matters in touch with each other, but no amount of care for references to formal pieces of work could sufficiently reflect the extent to which all her speeches and writing represented a legion of voices. John Todd, the ecologist, has designed a ship to be named for her, built as a wind-driven ecological hope ship that will move around the world providing various kinds of ecological first aid, picking up seeds and seedings of rare plants, particularly food plants, and propagating and growing them at sea so that they can be ready to plant or introduce as new crops on arrival in the next port, and this is what she did with ideas.

That notebook was a tremendously powerful tool, for consciously or unconsciously, we used to try to say things that would stimulate her to get it out and write them down, and then one would have the feeling of having contributed a piece to the complex jigsaw puzzle she was putting together. Later you might hear it in a technical lecture or read it recast as deceptively homespun advice in *Redbook*, lectures and articles that echoed hours and hours of conversation in a dozen cities. To incorporate even more people into the multilogue, she asked her audiences to write down their questions and would read afterward through all the questions there was no time to answer, to see what new chords she had struck with whatever idea or phrasing she was trying out.

Her visits to our households in Cambridge or Manila or Tehran were marathons of conversation, often exhausting. We sat and watched Vanni, as a baby, in her Jolly Jumper which allowed her to bounce and bounce, and I commented that one had to be careful not to leave her in the bouncer for too long since the gadget meant that bouncing was not self-limiting, there was no adult who would grow tired, no balance between adult patience and infant delight. That led us into talking about the difference between breast-feeding, where a balance develops between the infant's appetite and the mother's milk supply, and bottle-feeding . . . and we were off into a dozen different realms, television, industrialized agriculture . . . Always she was taking in, recombining, using and reusing every piece, bringing it out again in a new form and in a new place. When she was dying we tried to say things she would want to write down and keep.

I said to her when I was twenty or so and becoming involved in the pre-Vietnam peace movement that even if there were no nuclear weapons, we would still live with the possibility of building them, and that I saw that as potentially enhancing human dignity because it unifies the human species in a new way, making us responsible for each other and for the planet. The notebook came out. We talked about a friend, and I said, it's not so much that she enjoys suffering as that she suffers happiness. The notebook came out. Five years later, we would be talking of the same topic, and she would say, "That was your point, you know. That was a piece that you gave me."

Often it would be something I had forgotten and never thought of again, simply produced in conversation, which she had thriftily put away to recycle. Tiny increments, turns of phrase that seemed entirely trivial at the time, would be fruitful in combination with others. She spoke of ideas as if thought or knowledge existed in chunks that could be passed back and forth and collected: "We got an important point yesterday." "I got two or three really interesting pieces at the meetings; now I think I understand why . . ." "Have I told you that I finally got the point about . . . ?"

Many anthropologists today avoid writing things down

when they are with informants, either tape-recording to tran-
scribe later or making notes from memory, feeling that the
process of writing will interfere with their rapport, but Margaret used the notebook as a way of training people to give
her the kind of information she wanted. After a while they
would ask, "Why aren't you writing that down?" The best
informants would be people who, illiterate perhaps, were in
some sense intellectuals and would enjoy the process of
reflecting on and articulating their culture with the rhythm of
her pencil to encourage them.

Margaret had an insatiable appetite for the detailed and specific, for minute particulars. She grew up devouring nineteenth-
century novels, indeed all novels, involving herself in the
minutiae of thousands of fictional lives, and was much disappointed when I failed to respond to Dickens and Scott—and
to the details of the Bateson genealogy. She said, you'll never
be an anthropologist because you don't enjoy gossip, you're
not really interested in the details of other people's lives.
Later, when I became an anthropologist, I was glad to have
the models of linguistics and of my father's work, for I probably never will do the kind of classical ethnographic work
she had in mind.

Much of Margaret's appetite for detail came from a
desire to step into other frames of reference. Thus, she
wanted to know what I was thinking and learning, to get a
window into the world of nursery school or to know what it
felt like as a teen-ager to do volunteer work for the first
Stevenson campaign, to understand what Barkev and I
thought about Filipino politics, or to get the feel of Armenian
cooking. She wanted to perceive a slice of life, and then
quickly generate hypotheses, putative insights, "points," and
put them to work. When I was newly married and learning to

cook Middle Eastern food, she offered to cut up vegetables for a Syrian salad and after I had said, no, not that way, about the tomatoes, and then again about the cucumbers and green peppers, she drew the abstraction that the rule was that all the vegetables should be in comparable-size pieces, not some chopped and others sliced, and went around the country quoting the rule with a pleased sense of insight.

Sometimes Margaret would announce that she wanted to interview me on some issue, and this was a signal that she wanted rich description, not just opinion—a narrative of how such issues as holidays were handled at the Downtown Community School that might explain an emerging tension between Jewish and Gentile parents, for instance, or enough details about the scuffles that sometimes broke out after school with neighborhood children to make specific recommendations.

Sometimes she sent me to interview someone else on a subject I was interested in that she did not feel sufficiently informed on, taking notes carefully each time. At ten I became interested in raising money for the cancer fund and she sent me to interview Dorothy Lee, an anthropological colleague, who had had a mastectomy and recovered. When I started to keep house she suggested that I interview her about cooking on a limited budget as she had had to do when she was first married.

She both read and wrote extremely rapidly, setting herself time frames and quotas that she would find satisfaction in exceeding. Like Gregory, I have tended in some periods to limit what I want to take in. Gregory, indeed, read very little, no more than two or three books a year that really entered into his thinking, almost as if he were protecting his capacity to concentrate. The list of books that Margaret coaxed into being from colleagues who needed help in crystallizing their ideas would be a longer list than the list of scientific books Gregory read with care.

Margaret's networks included people of all ages. For many years, she was the youngest person in most of the groups that she worked with. Then she realized that she was continuing to work with the same senior scientists and a few peers, all growing older, and no younger people were being included, so she undertook an ongoing battle to get younger people brought into the discussion.

She had a strong belief that the experiences of children born after World War II were so different that they were almost like members of a different culture. She included me in the postwar generation because of the curious accelerated atmosphere in which I grew up, even though I graduated from college in 1960, ahead of the generation of students who came in with a shared sense of dissonance so strong that institutions were shaken to their roots. Thus, she was certain that I and my peers, and the students in her classes over the years, knew things she did not know and needed to learn. What is it like to live in a world that has *always* been under a nuclear sword? What is it like to live in a world in which war—the kind of war that the members of her generation were so dedicated to winning—is obsolete? Cross-generational conflict peaked when the first generation that had lived in this context became young adults, and calmed down as they graduated and moved into the hierarchies and decision-making positions.

Whenever a group project, like a conference or a committee report, was being planned, she would pose the question of how to get young people included. This is a good deal more difficult than it would seem, because even if one overcomes the status consciousness of senior participants and the notion that travel money would be better spent bringing one more expert, it was always difficult to say which young people, to represent a group in which individuals had not had the opportunity to show their potential stature. She tried a dozen different models, such as drawing on the local student population wherever a meeting was being held, or suggesting that each senior participant be able to designate one junior, and often she fell back on including a member of her family—it turned out that if you were famous you could be indulged in the desire to bring along a daughter or a godson or a niece, when a principled insistence that younger people be included was often ignored. So for years she would fall back on inviting me or my cousins or the children of close friends as a personal prerogative, to make sure that someone under thirty would be present and participating. Because she was concerned with historical age, with where people stood in relation to the generation gap, she continued to call on several of us as we passed that chronological line. Three or four people have said to me, "You know, when you were out of the country, your mother treated me as a kind of substitute daughter." One reason for this was that in this way she could make sure certain points of view were included, certain voices heard.

The most uncomfortable occasion of this kind was a conference at Georgetown University on language teaching, which was to take place without any comment from language students although there were some demonstration classes. Exasperated, my mother got me invited to be on the panel as a language student who had been studying one Middle Eastern language after another, but in fact I was also at the time a graduate student in linguistics, slipping awkwardly in and out of the role of a linguistically naive language student, as uncomfortable as an adolescent boy who doesn't know from moment to moment whether his voice will come out soprano or tenor. Mostly, however, she was not seeking someone to act out the role of youth but to embody it, assuming real differences in experience.

In the wartime and postwar studies of contemporary cultures, when it was impossible to do fieldwork on the spot because of war or Cold War, groups of people were brought together who replicated among themselves some of the diversity of the culture under study, who could represent not only classes of individuals but, in their relationships with each other, the relationships between the sexes or the generations or between different ethnic groups in the population. When Margaret put together a panel and tried to make sure of the inclusion of members of different groups, she was not concerned with having "one of each" but with being sure that the group reflected certain kinds of relationships, certain inevitable contrasts of approach. Still, no one was there as a one-dimensional token. Generally speaking, the dimensions of counterpoint were multiplied, so that, for instance, when she did a book with James Baldwin based on a recorded dialogue between them, they evoked the sense of multiple dyads challenging each other with different kinds of wisdom, a man and a woman, black and white, a poet and an anthropologist.

As a result, our first professional collaborations were when I was brought into projects to represent childhood or youth. It's important to understand that she was not interested in a statistical kind of representation, or in finding the least common denominator of a particular part of the population, for anthropologists try on the whole to deal with the uniqueness and idiosyncrasy of each informant, and value insight and articulateness. Once she brought home a test that had been developed to be used on new army inductees, based on a series of questions about how a puppy would behave when faced with certain situations that represented moral dilemmas, and she tried it out on me. With excessive sophistication, I asked if she wanted me to respond as if I were a dog or as if I were a GI or as if I were myself, and she made me run through all three sets of answers. She would send colleagues and students to interview me, especially those who were just learning how to interview. I produced dozens of dirty jokes for a researcher on children's humor, and long reflections for a study of a child's morality. (What is the worst thing a child could do? Destroy the world, I said.)

It seems to me important, looking back at these relationships across a decade and a half when I had assumed the professional identity of an anthropologist and became a colleague in a new sense, to see all of these as essentially collegial relationships. Even as a very small child, I brought expertise on a particular part of the culture and might make comments that would contribute a key ingredient to some larger understanding, comments that would be treated individually and not statistically, even though I was a pretty unrepresentative child. An adolescent girl in Samoa, the little boys who worked for her in Manus, her deeply sophisticated and skilled secretary, Made Kaler, in Bali—each of these might supply a crucial angle of insight on a culture or a generation because each spoke from a distinctive place within it, so she listened simultaneously to the individual and to the cultural themes that individual was expressing.

She once said in some exasperation that I seemed to want to do exactly the same things as the rest of my generation, but with more elegant and sophisticated reasons why I needed a car or why I planned to marry at twenty. I remember that comment often when male friends explain to me that the reasons why they have walked out on a marriage have nothing to do with the fact that half a generation of their peers are busy doing the same thing. The fact that an individual is acting within a cultural pattern is no reason for disallowing his or her sense of decision or urgency. She felt strongly too about not using a knowledge of cultural pattern to manipulate.

Much of what is recounted in this book about others and even about me is here because of her comments, because of the ways in which she shaped observation or memory. What I said when the first atomic bombs were dropped comes not from my memory but from hers and is significant because of what it tells of her way of observing. As I write, I find myself reluctant to sort and judge, weeding out the trivial and underlining major points, because it is the varying texture that I most want to convey.

There is no way, I believe, that Margaret's work can be studied to winkle out the influences of different people, no way the composite can be deconstructed, as literary theorists say now. But the task can perhaps be turned on its head if we assert that the greatest part of her originality lay in the invention of ways of listening and synthesizing ideas. By her peripatetic life, she created a kind of multilogue to which individuals contributed who would otherwise have been in isolation, their ideas never juxtaposed in fruitful ways. There was a change of emphasis as the number of partners increased, since the kind of intensive and sustained partnerships central to the Balinese and New Guinea fieldwork decreased, yet even to the end of her life one could make a long list of books that would never have been written except for the infusions of energy she gave, the insights she elicited. She was at her best as a visitor in other people's intellectual territory or tasting their experience, not demanding that they share in the breadth of her own. It was a sign of fatigue and, eventually, of age, when she began to rely increasingly on preformed material, able to maintain a witty discourse that stimulated admiration when she was too tired for careful listening.

Some of Margaret's finest and least-known work deals with issues of the nature of the discourse and how it allows ideas and the common cultural heritage to evolve. Indeed, it would be helpful if some young scholar were to approach her work as a sociology of knowledge. The key work is the volume of her Terry lectures at Yale, called *Continuities in Cultural Evolution*, where she talks about evolutionary clusters—the groups of people among whom ideas develop and within which the contributions of an outstanding mind resonate and are amplified: the twelve apostles, Paliau's immediate supporters, the Manhattan Project, the Macy Conferences on Cybernetics. In the same volume she also talks about the means of transmission of new ideas in visual symbols or even with toys that can be held and handled, as well as with words, and about how ideas can be protected from caricature and oversimplification on the one hand and from sectarianism on the other. She also wrote a book about conferences, transient clusters, and about the use of groups of research. Arguably, *Culture and Commitment*, which deals with communication across and within generations, could be seen as part of the same set of writings: the process of change which she had wanted to study on her first field trip when Boas pressed her to study adolescence, the modalities of change in which she was herself most active.

When I got my Ph.D., Margaret wrote to congratulate, addressing me as *"Chère Collègue."* Often indeed she would sign herself as "Margaret" or even "Margaret Mead" instead of "Mummy," and then quickly add *"sic"* to show that she had noticed. The conversation and exchange were so rich that I find my thinking saturated with hers, and have great difficulty if I try to untangle who said what. The problem of specifying her contribution is increased by the fact that it is not a single structure of formally related ideas, a system of thought like my father's, but more "a way of seeing."

Intense conversation did not by any means consist of agreement only. I grew up accustomed to taking the report "I had an argument with so-and-so" as a report of exciting and stimulating debate, not of conflict, and she would go at a discussion hammer and tongs, expressing exasperation and disagreement: "I never heard such rubbish" or, famous now from people's reminiscences, "fiddlesticks!" She and I had an argument at a Russian Easter party once about concepts of leadership, and then laughed together on the way home at the way others had dropped timidly out of the vehement conversation and the possibility that they might feel they had witnessed a quarrel or a breach. And even in a few relationships where discussion was sometimes infused with emotion to the point of real quarreling, she would return to an affirmation of how much she got from the stimulation of the other mind, recognizing the intrusion of other layers of personality and experience but refusing to disengage because she felt she was learning and sharpening her ideas all the time. Indeed, she sometimes said of a marriage that seemed to be headed toward dissolution, "They don't even bother to fight anymore."

Talk was a passion. Margaret once described to me how

one of her colleagues had come to her hotel room at night, determined to make love, and she, not wishing to reject him, had welcomed him and held him not in physical embrace but in impassioned conversation about his work until he departed in the small hours of the morning, all passion spent. That seems to have been the pattern for long periods at the beginning of Margaret's relationships with Reo and then with Gregory, and during her long engagement to Luther, and it sometimes strikes me as I talk to people about her that many of them are not entirely sure themselves exactly what happened, how the intensity of their relationship was expressed. The metaphor of lovemaking and conversation recurs repeatedly in her writing.

So much talk over so many years meant that a sort of code could develop. In conversation with Margaret I could allude to a person or event, the name forgotten, and she would complete the reference: "You know, the woman in Istanbul . . . the man who talked about the two-headed turtle . . . the man who came to Cloverly and never took his jacket and tie off . . . the family who had that Bateson portrait . . ." She seemed able to identify people not only by the detail mentioned in a description but by what the choice of a particular point of identification said about the total gestalt. The other feature of our conversation that gave it a codelike quality was the way we used to complete quotations for each other. One would begin the quotation of a poem and the other would join in to recite in unison. Edna St. Vincent Millay, whose poem "The Blue Flag in the Bog" underlay all of Margaret's discussion of nuclear danger, turned up in a lighter mode as well, expressing the headlong embrace of life. "My candle burns at both ends; / It will not last the night;" and then both of us together, "But, ah my foes, and, oh, my friends—/ It gives a lovely light,"[1] or again "We were very tired, we were very merry—/We had gone back and forth all night upon the ferry . . ." and then together, "And she wept, 'God bless you!' for the apples and pears, / And we gave her all our money but our subway fares."[2]

All her prose echoes with lines of memorized poetry, referred to sometimes with a single word or phrase, the work of the poets who were her contemporaries and also the classics of her childhood, echoing more strongly than what she read in later years: "Over the river and through the wood . . ." "Out of the night that covers me . . ." "and floated down to Camelot . . ." The words of the Bible and the Anglican liturgy echo as well: "We have left undone those things which we ought to have done. . . ." "The peace of God which passeth all understanding . . ." "My cup runneth over. . . ." "All things work together for good. . . ." A single phrase would cross-index a dozen previous discussions, a reference to the parable of the talents for instance, which turned up repeatedly in her thinking, or Kipling's depiction of heaven as a place of tireless artistic creativity:

> But each for the joy of the working, and
>   each, in his separate star,
> Shall draw the Thing as he sees It for the
>   God of Things as They are![3]

There was nothing of pretension in this play, no ostentatious erudition as in certain styles of Shakespearean quotation, since indeed many of the things she quoted are out of fashion now, the old chestnuts of popular culture, but part of the furnishings of our minds. Echoes of such familiar phrases are what give us a lump in the throat when we read them, effective precisely because they are clichés, because they evoke emotion associated with past uses. Margaret once commented on the work of a poet friend that he used a vocabulary almost entirely of words of Anglo-Saxon rather than Romance origin. In many passages of her work she does the same thing, evoking the gospel with references to women "great with child" rather than pregnant.

The deliberate use of such evocative language in her writing goes back to the decision to write about Samoa not only in a way which would be intelligible to non-experts and especially to teachers, but to make it possible for readers to respond emotionally as well as intellectually, to get a "feel" for the quiet backwater of Samoa in which she worked, a mood that might be interrupted and has certainly changed over time, but that was the pervasive background of life as she experienced it herself. She is no more hesitant in saying that the night is "populous with ghosts" than in saying that the sun rises. She says, "the old men sit apart, unceasingly twisting palm husk on their bare thighs and muttering old tales under their breath,"[4] instead of saying something like the following: "Approximately 80 percent of the older men (defined) spend two or more hours a day twisting palm husks. The ethnographer noted that several commonly mutter inaudibly while engaged in this task and surmised, on the basis of interviews at other times when folkloric texts were elicited from them, that they might possibly be engaged in reviewing or rehearsing narrative materials." Her sentence conveys something of the partial withdrawal of the old, their involvement in muted activity rather than quiescence, their role in the preservation of tradition. At the same time, it invites the reader to share Margaret's experience, which has an aesthetic unity, even as she paralleled that experience with the detailed recording of times and places and names, all available for further study, not in the swift personal style of her notebooks but typed and legible, a public property.

She chided me, when she read my drafts, for the way in which I drifted into the use of typical academic hedges: If a statement is not a statistical statement one should not protect oneself from a statistical misreading—no need to assume that a reader will read "the old men" as "100 percent of the male population over age sixty-five" or "sit apart" as "spend twenty-four hours a day in a position defined by . . ." There is more than one kind of generalization. Truth is not served by the self-defensive peppering of a text with adverbs like "apparently," to both affirm and deny the self-evident limits on observation.

Margaret put a great deal of thought into matters of phrasing, knowing how this would affect responses, blaming herself when a phrasing was wrong. Thus, a tremendous flap was created when she proposed the "legalization" of marijuana, and she felt that it was her fault for not thinking through the fact that the issue, for marijuana as for abortion, was one of decriminalization. Similarly, when, after she said to me as a child that "the white man" had taken the land away from the Indians, I said that *I* hadn't taken it away, she blamed herself for conveying a concept of shared guilt that she did not believe in, instead of saying, for instance, that the early settlers had taken the land.

A phrase, a fraction of a line that evokes a whole range of associations and ideas, this way of proceeding is congruent with her sense that you could experience the whole of something from just a little sample. She carried with her, all over the world, a small silk pillow that would allow her to sleep anywhere and evoke the comfort of a bed at home, and she would refer to a very brief experience as explaining whole worlds in which others lived: "Just once when you were a baby and wouldn't stop crying I felt I knew how mothers feel who batter their infants." "Once in New Guinea I was depressed and knew how it would feel to go through life in a permanent depression." During one week when I was a teenager we had several arguments from which I stormed out in tears, and later she said, "You only acted like an adolescent for a week but it was enough to make me think about sending you off to boarding school, so I know how other parents feel."

As I grew older and became more of a colleague, I began to observe more systematically the way she worked, comparing her advice on public speaking with her actual performances and studying her comportment. She advised me on how to sit on a platform without slumping—you get one foot behind the other before you sit down—how to focus on one or two faces in the audience, looking at least one real person in the eye, a technique she learned from her father. One had to hear many of her speeches over time to understand her technique. She almost always spoke extemporaneously but wove together old materials like the "singer of tales" who weaves and reweaves familiar materials and phrasings of folk epic into new extemporaneous renderings, or the jazz musician engaged in practiced and familiar improvisation, moving by organic connections and associations from one comment to another to cover the rough outline she had blocked out in advance, touching on a few new points of unfolding thought and curiosity.

We were together at half a dozen conferences or ongoing seminars. The first was the 1962 Conference on Paralanguage and Kinesics[5]—for which I was rapporteur at the suggestion of my Arabic professor, Charles Ferguson, who was a key figure in the early development of sociolinguistics— where she made the final synthesizing speech and proposed calling the discipline semiotics, in the plural. There were many others, at some of which our meetings were coincidental while others were contrived by her. I used to sit there, paying as much attention to her silences as to her interventions, for I knew her thought well enough to notice the occasions when she would have had something to say had she not been saving her interventions for major issues or for opportunities to synthesize. Often I was amazed at her self-restraint.

After I had worked with Gregory on three or four different conferences and published *Our Own Metaphor* about the first, she sat in my New Hampshire kitchen and said, "I think you and I ought to have a conference too; what shall we have it about?" and I suggested ritual. As we explored the idea, tossing it back and forth over several long country evenings between household tasks and caring for Vanni, it turned out that both of us were really as interested in the process whereby new rituals are formed, ratifying social change, as we were in descriptions of stable ritual systems.

By that time we had had long discussions of the work of the Liturgical Commission of the Episcopal Church in which we were both more radical and more conservative than the church—we wanted the retention of the old along with the new and we also wanted ways to bless all significant transitions: new unions, including homosexual ones, ways to mark divorce as well as marriage, and ways to mark moves from one community to another. Vanni had been baptized in an experimental rite combining the three sacraments of baptism, confirmation, and communion, which was eventually not adopted by the church, and I wrote the prayers myself, as I had written texts for other kinds of experimental services. Margaret believed strongly in the need for new symbolic and ritual acts to focus and intensify concern, and was involved with the beginning of Earth Day and the notion of ringing the United Nations Peace Bell at the moment of the vernal equinox.

In the end we applied for funds for a conference titled Ritual and Social Change, one more conference in the castle of Burg Wartenstein, a conference full of tantalizing pieces that finally failed to jell. Margaret felt that it failed because we had combined people who wrote about religious phenomena from the outside with those who were also committed and engaged, the scholars unwilling to allow the expression of experience, religion being more embarrassing to talk about even than sexuality. Feachan O'Doherty, an Irish priest-psychologist who had worked with Margaret in the World Federation for Mental Health, said mass each morning in the old tower chapel of the fortress, and members of the group hesitated between interest and embarrassment, engagement and cynicism.

There were other times of drawing me into her work when the professional and the personal, the objective and the subjective, were so entwined that it seemed impossible to separate them. She would offer herself as an informant for one of her classes in field methods as they learned to interview her on some aspect of her experience, and then ask Vanni and me to come for a practice session on observing and recording mother-child interaction, while one of the photographs of the session became the family Christmas card.

Once in a while, however, there was an occasion when sharper separations seemed important. Throughout her life, her professional relationships with doctors or lawyers were also personal ones; she regarded the monitoring of her own sensory skills and the exploration of her mental life as both a professional obligation—understanding the central tool of one's work—and a matter of personal fascination. Once when her physician remarked that unlike most women of her generation, she did not bring a handkerchief with her when she went into the examining room, she replied that others bring a handkerchief to retain a shred of the personal and social in an impersonal environment, whereas she left the hankie (she who almost always had one) behind to depersonalize and formalize an already personal relationship. She was disconcerted when she asked to see her record and found herself described as an "obese white female," and then, amused, told the story on herself for weeks, not only observing herself but observing herself being observed.

XIV
Steps to Death

Today we no longer live in a culture dominated by the concern for a "good death." Indeed, we are so preoccupied with fending death off that even though the last few years have seen a burgeoning of discussion of how to meet death more honestly and simply, families and individuals have immense difficulty finding their way step by step into the expression of shared feeling. The timing of death, like the ending of a story, gives a changed meaning to what preceded it. Death is also the moment of a gift that the old give to the young, a last opportunity to teach about life, but it is so shaped by accident and happenstance that the moment may pass and wisdom is often unexpressed. The young stand by, wanting to give back to their parents some fraction of the care given to them, hoping that their parents will meet death in a way that softens the anticipation of death in their own lives and reconfirms their sense of the integrity or generosity of their parents.

By the time we are adults we have said a great many good-byes, often knowing long periods of minimal contact with parents as I did in childhood when they were away during the war, then later, when their separation from each other meant they were in different and remote places, and finally as I went off and spent periods of several years in other countries. When parents die, all of the partings of the past are re-evoked with the realization that this time they will not return, and the distances of separation, that for me spanned round the globe, become infinite. And yet the familiar experience of reunion persists, expected beyond the accustomed separation.

In 1978 both Margaret and Gregory were diagnosed as having cancer. She died in the fall of 1978 and Gregory in the summer of 1980. In the spring of 1978 he had come close to death, initially seeming far more threatened than she, but then had a remission. For me, the two deaths fall in a single period, signaling a time of transition in my life when other things were changing rapidly also, as the career that had kept my husband and me in Iran for some seven years collapsed in the Iranian revolution. The deaths of parents make all the dress rehearsals of adult independence real; for me they coincided with the necessity for new beginnings.

One of the systems of ideas that my mother felt closest to, which provided the context for much of her thinking about childhood and about families, was Erik Erikson's analysis of the human life cycle. Erik's way of thinking has been important to me as well, for the family friendship with the Eriksons opened the way for me to teach in Erik's course when I was on the Harvard faculty. My mother used to scold Erik, when he first arranged the developmental stages on a chart, insisting on the importance of making the sequence of a life read upward and not downward. During the years of the "generation gap," Margaret developed a theory of changing modes of cross-generational communication, building on age-old harmonies in which the patterns echo each other, each generation to some degree prefiguring the next, and provide solace for the old and hope for the young.

One of the tasks that is still to be done is to work out the consequences for individuals when life cycles overlap in different ways, when a particular rhythm of overlap is conventionalized in a society. As my mother's death approached, I asked myself, what difference would it make to me if I were twenty or thirty . . . or fifty or sixty years old? And what difference would this make to her? Today we have people who retire from their careers at sixty-five to care for aging parents, and when those parents die their children must carry the grief without a sense of vistas of possibility still ahead of themselves; aging parents end their lives seeing that in another, less literal sense, their children's lives have ended as well.

In the modern period, the life cycle has been stretched, but far more than it has been stretched, it has been extended. Although childhood is longer, marriage later, and childbearing viable until a later date, the greatest change in the life cycle is in the increase in the number of years lived after childbearing. It used to be true, and still is where nutrition and public health have not been greatly altered by modern conditions, that although most people could hope to see at least one grandchild, few could hope to see a great-grandchild. In effect, they could expect to die while their children were still engaged in raising the next generation, involved in needed caring and with so much still unfolding. There would be the question of what to tell the children, whether to speak of death in whispers and send them away or to involve them in ceremonies; there would be a knee baby to be hugged, the insistent needs of a young family for food and attention to override immobilizing grief. My sister-in-law who went into labor with the first grandchild on the day her husband's father was buried, the boy who rises to say Kaddish for a grandparent in the same year in which he becomes *bar mitz-vah* . . . these express the range of an ancient rhythm. When my husband and I went to Beirut for the first time after Vanni was born, to show her to his family, we took her, aged two, to visit Barkev's Aunt Lucy who had never married but instead had cared for him as a child. She talked and laughed with us and then, having had that long awaited meeting, she stopped speaking and quietly died within a fortnight, the rhythm of ninety years complete.

If my mother and I had had children quickly when we got married, I would have been over fifty when my parents died, and Vanni thirty. Instead, I was approaching forty and Vanni was not yet ten, old enough to weep and then go out to play, almost the same age as I was when my grandmother, Emily Fogg Mead, died; I can celebrate the fact that Vanni was old enough to know and remember these two famous grandparents whom she will go on meeting in print and in images fashioned by others all her life, old enough to have looked at tide pools with Gregory and to have asked her grandmother to speak to her kindergarten class because the other children wanted to hear about "cannonballs." So many endings came together for me that it was several years before I realized that my age at the time of their death was the same as theirs at the end of the war—that the images of them that fill these pages are images of them older and further into their lives than they will ever know me to be.

I had seen my mother in the fall of 1977 at the meetings of the American Anthropological Association in Texas, where she moved with authority through the crowds with her thumbstick and scarlet cape, known and recognized by everyone, and treated with awe as a link with the history of the discipline. In the course of her lifetime, anthropology changed from the preoccupation of a small group, all of whom knew each other and felt that there were insufficient hands for an urgent task, to the business of a vast association of strangers, many of them searching for jobs in an over-loaded profession. We talked about my own plans and she complained that she was having to be very careful of what she ate because of recurrent attacks of diverticulitis. She had already begun to lose weight.

Back in Iran, in the spring, I got a phone call from her saying that Gregory was in the hospital and had had exploratory surgery for a cancer of the lung that was found to be inoperable. The assumption was that he had very little time left, so he had called Margaret from the hospital to dis-cuss whether it would be appropriate to ask me to come back from Iran to help him with his uncompleted book. Gregory had gone very rapidly right to the threshold of death and then was battered by the attempt to determine a course of treatment. Three times in the hospital they had him in inten-sive care, trying to deal with pathologies created by diagnos-tic efforts.

After I arrived, as Gregory recuperated from the indigni-ties of modern medicine, it became clear that he was recover-ing ground lost as long as a year before diagnosis and was into a long-term remission. In the hospital he was realistic, interested, and philosophical, insisting that the inability to face death is one of the major flaws in Western civilization, and confronting mortality with one eyebrow cocked and a quizzical expression. He lectured the doctors and nurses and also the governor, Jerry Brown, who came and sat by his bed-side, on the nature of social and ecological process. A healer who visited the hospital room the day after his surgery announced that the cancer was dead—there was nothing more wrong—but did not claim to have brought about the remission. Others have suggested that the remission might have had something to do with the fever accompanying a severe bout of pneumonia early in his hospital stay. He commented that the medical profession dealt magnificently with the worst emergencies and seemed to keep creating them. At the same time, although he was profoundly convinced of the reality of the natural processes emphasized by the holistic health movement, he was skeptical of most of what they proposed.

I arrived the day before Gregory was released from the hospital. As soon as he returned to the family house in Ben Lomond near the Santa Cruz campus of the University of California where he had been teaching for several years, we developed a routine of work, with the entire family organized to support him in his desire to complete the book that would relate his thinking to problems of evolution, thus finally completing his dialogue with his own father. He had a stack of transcripts of lectures he had given since the publication in 1972 of *Steps to an Ecology of Mind*, a book that anthologized almost all his published articles and allowed him to see his diverse ventures into the study of primitive cultures and schizophrenics and dolphins as all part of a single exploration. After it appeared, he had quickly become well known and indeed something of a cult figure to the ecology movement as well as to California New Age communities, and had begun lecturing around the country. The need to speak to more general audiences, in addition to his role in organizing a series of conferences, had moved him toward a new sense of what he wanted to say about the nature of mind and the interconnections among all life.

He would wake up at two or three in the morning and I would hear him on his way down the hall. Then we would heat up oatmeal and make strong coffee and start working on the manuscript. He talked into a tape recorder, on the great oak dining table, with periodic breaks when we had long discussions of the concepts he was feeling his way toward, and then the next day I would type what had been done, finding places that needed to be amplified or clarified in the next night's work. Usually we managed one work session together in the daytime as well, with sleep spaced out in between and family meals with Nora and Lois and short forays outside the house to walk or sit in the sun. Together we sat on the veranda, watching peacocks in elaborate courtship roaming loose on the estate where the Bateson house was located, and we tied the observation in with the questions of symmetrical and complementary relationships that he had played with for years.

By the end of a month, with Gregory visibly mending every day, the newly composed passages were sufficiently melded together with the revised transcripts of lectures that a book was nearly completed. Friends were still writing him letters of condolence on his impending death. He was moved by some and stung by others to write back irritably, chiding, you never have really understood my work. The final chapters were completed after my departure, *Mind and Nature* went to press, and almost immediately Gregory signed a contract for yet another book. The family moved from the house in Ben Lomond to the Esalen Institute where he slipped into the role of resident guru and skeptic, lecturing occasionally, until his final illness.

When Margaret heard of Gregory's cancer, she had already been told that the pain she associated with diverticulitis, by now giving her great difficulty eating and causing attacks of pain at all hours, was in fact an inoperable cancer of the pancreas. She also had a spot on her lung. In an odd way, she enjoyed that metastasis on her lung, feeling that it was a token of some deep sympathy with Gregory's lung cancer, but she resisted acknowledging the cancer of the pancreas, preferring to listen to a healer who insisted it was a less serious cancer of the intestine. She acted on the theory that Gregory's illness was the serious one, only sending me to her physician to hear about her own condition when I passed through New York on my way back to Iran from California. She went in and out of focus in willingness to deal with what was happening, planning her activities for a five-year span into the future.

Death is something we were used to discussing, or so I thought. The first person I knew to die was Ruth Benedict in 1948. Although Ruth had completed her doctorate in anthropology only a few years before my mother, this was already after several different abortive caeers. She was sixty-one when she died, wearied by hectic effort during the war. I remember her in death as archetypically beautiful and cold, her large luminous eyes closed, her white hair perfectly sculptured. My mother wrote, "When she died, she looked incredibly old, as if the wisdom and suffering of several hundred years was momentarily expressed in a face which, for that instant, seemed more than life-sized. She had always felt so strongly about the beauty of the dead, and we brought our children to see her, giving them a protection which few children have today, in an acceptance that death is a part of life."[1] In a real sense, I faced that year of death and rumors of death across Ruth Benedict's coffin. Only a few years after Aunt Ruth's death, my mother's mother died, after a series of strokes, giving me forever the sense that as people approach death they become smaller and smaller.

Margaret had given a good deal of thought to the mode and handling of death. In the mid-fifties she was among the first to draft a document designed to protect her from medical interventions that would keep her alive beyond the point where she ceased to be truly herself. Her instructions were considerably beyond the form the discussion has taken in recent years for she made it clear that she did not want to live with any reduction in mental capacity or with any combination of impairments of her ability to communicate and move around.

I argued vehemently as a teen-ager against this anchoring of the meaning of her existence in the ability to function at top intellectual capacity. Some qualities recede or fade more quickly than others, but I felt—I still feel—that although it is right for an athlete, in her youth, to organize her life around her fleetness or her agility, it would be wrong to take these as the chief and only value of life, dismissing the potential meaning of many later years and in a sense dismissing the value of those others who live out their entire lives slowly and clumsily. There are people who respond lovingly to those around them long after the skills they claimed in their prime are past, and it is not a bad thing to discover that one is still cherished when no longer quick or witty or beautiful. As a teen-ager, not knowing what my own life would prove to be about, I suppose I was threatened by this implicit statement of what my mother felt was central to her own life—thinking, communicating, grappling with problems, moving around; she would not simply stay to be loved.

Margaret speculated that the present period, in which many elderly people become stranded on shoals of unsatisfactory survival, would be only temporary. After all, she pointed out, like most people of her generation, she had been near death from illness several times as she was growing up, with diphtheria and influenza and scarlet fever. Her body had perhaps learned to mobilize for survival under conditions of high fever, delirium, or near coma, acquiring a certain stubbornness that would be hard to lay aside later. I, on the other hand, in the era of antibiotics and immunizations, had had no such experiences and so no deep learning to cling to life. Perhaps people of my generation will finally hold rather lightly to life as our bodies weaken, and simply let go as they say some prisoners of war have done in the post–World War II era.

When mortal illness finally arrived, as Margaret became sicker and sicker and dogged by increasing pain, she became more and more determined to resist death. An old rivalry with Gregory reasserted itself in her conviction that if he could beat cancer, so could she. There was almost the sense of a dare, based at a deep level on the notion that a mind of sufficient strength and focus could defeat the disease.

This is what she was saying in that brief interval in New York, when I returned from my conversation with her doctor, insisting that she could recover from cancer by working with the healer, who had assured her that it was really in the intestine. I knew that she had decided against all chemotherapy or other intrusive therapies, but the doctor had responded to my questioning by saying that I should not feel obligated to press her to accept such treatments as the responsible course, since he believed the cancer was terminal in any case. Back at the apartment which she shared with Rhoda Metraux, she spoke of her intention to set up a small independent household, and I wondered whether to stay with her for whatever time remained, but she insisted on my going. She was clearly unwilling to accept anyone by her who would not abide by her conviction that she would be healed; indeed, she needed my departure for Iran, where I was working on the development of a new university, as a reassurance. "Look at her," she boasted as I got ready to leave for the airport in a new dress, "doesn't she look too young to be a dean?"

For the summer of 1978, Margaret organized a conference on the future, at Chautauqua, that would bring both Gregory and me together with her as colleagues for the first and only time, along with several others she had selected. Each of us gave a public talk to the Chautauqua audiences and Margaret and Gregory sat together like old lovers reconciled while I spoke in that vast auditorium about attention, each acknowledging in me a maturity I had brought back from Iran. In between the lectures, we met to talk and think together. She believed, I think, that the group she had chosen might in some way give birth to a crucial new idea.

She saw the eighties as a period of immense and increasing danger as economic disparities widen and the partial and inadequate mechanisms that restrained the use of nuclear weapons during the first thirty years of their existence are eroded by proliferation and changing technologies, and she wanted to draw on metaphors from fields as disparate as cancer research and the study of ritual to think about what could be done. We were all of us so preoccupied by the sense of her pain that I have great difficulty remembering the discussions. It is true that sometimes the image of one suffering human being, a napalmed Vietnamese child held by his mother, Christ on the cross, a starving infant in the Sahel, can enable us to respond humanely and imaginatively to disasters affecting hundreds of thousands, but these are images only, opening out to broader understanding, instead of immediate points of focused experience.

I shared her room, getting up through the night to bring her tea or ice packs when she thought it would help or simply holding her, small and soft and scented with sleep, filling my mind with memories of coming upstairs at night as a child and climbing into her bed. Yet each day she pulled herself together for her public performances, winning and holding the audiences and answering questions with rapid wit, rationing her energy and the supply of pain-killers. She had sprained an ankle and was in a wheelchair that week, without the strength to use crutches except very briefly, and I went with her on and off her lecture platforms and helped her make improvisations of clothing that would not hang too loosely. Through that summer she continued her professional traveling, with different people nursing her from one public occasion to another, and even took a last trip to lecture in California where she had further hours of talk with Gregory.

In the fall I was back in Iran when she entered the hospital, insisting it was only for a few days' rest. By this time she had changed her physician twice, unwilling to accept any negative prognosis or denial of the efficacy of the healer's treatments. Friends telephoned me in Iran and suggested that this might be a last chance to see her, but she was angry when she heard that I was coming. She was concerned that my absence during that period of approaching revolution might imperil the possibility of remaining, since both Barkev and I were hoping that our work would be possible in a postrevolutionary Iran.

When Vanni and I arrived in New York, we went straight to the hospital. We came into the room and she awoke and saw us and sat up, immediately onstage, and showed Vanni the lights of the river through her window. The next day, when Vanni was not with me, she said, "You know, I've been talking all these years about the need to expose children to death, but it never occurred to me before that they need to see suffering too, as you are showing Vanni." She went on to talk about societies in which children of Vanni's age are not only responsible for younger siblings while mothers are preoccupied with new infants or work far from the house, but are also responsible for the care of the aged and infirm, so that often a small child is the only one present at the moment of a stroke or a heart attack.

We were there for two weeks, a painful and disturbing time, not just because of the approach of death but because of Margaret's loss of the ability to orchestrate what went on around her and her insistence that we continue to speak as if she were going to recover. I once read a fantasy in which a sorcerer had sent out replicas of himself to do his will on different continents, but in the end his resolution and the power whereby he sustained them in far places weakened and they returned, coming remorselessly through the rainy night, in a fatal convergence and reunification. Around my mother's hospital bed, conflicts flared up between people who had been important in different parts of her life but had no friendship for each other, insisting competitively on the right to a portion of her time and attention, overwhelming any efforts to limit the flow of visitors. Again and again, even when she was drugged, she would respond to different visitors with the specificity and warmth they needed, like an aging vaudeville star pulling herself together for one more bow and smile.

Enacted in front of us was the fragmentation she had written about in 1955 in the letter prepared to be read after her death, as different and fragmented relationships clashed and she was continually oppressed by the fear of being left alone. She had private nurses at all times and almost always some second person there so she would not be alone for even a few minutes if the nurse went out to make tea or fill a hot-water bottle, and yet the management of these comings and goings was a continual conundrum of missionaries and cannibals around the bed and in the hallway.

Margaret pressed me to depart and go back to my work, denying her own illness. I wonder now whether some of her insistence came from a recognition that the impulse to be present beside a deathbed is a repetition of the attraction of the cradle. Women in our society expect to have to resist pressure to put aside their careers and devote themselves to their children, but are taken by surprise by the pressure to dedicate themselves to the old, especially during months or years of illness. Such pressure, which would not be put on a man, implies that a continuation of professional commitment is a frivolity in a woman, a matter of selfish choice. When Margaret spoke of women torn between work and child care, she quoted Harriet Beecher Stowe as saying that she was distracted from her novel because "the baby cries so much," but she pointed out that the real problem is that the baby smiles so much, that the care of an infant is a perfection of creativity and delight that few can achieve in any other kind of work. The task of helping another die is equally a matter of moment and finality, a task whose significance dwarfs others into busy work, and yet we judge lives not by their beginnings and endings but by what has been done in between.

I think that the firmness with which she urged me to leave and return to Iran, where my own work was in crisis, was a reaffirmation of choices she had made throughout her life, particularly as she carried on her work through my childhood. It was more than generosity to Gregory, then, that made her agree that it would be proper to ask me to come when we believed he was dying. Gregory was asking me to come and help him complete his book, a vocation beyond that of care and tenderness in a life in which "having a book together," creating something permanent beyond the transient repetitions of procreation and mortality, was the ultimate expression of intimacy.

When it was time to leave, I tried once again to share an acknowledgment that this was probably good-bye, but this was still not acceptable to her. A fortnight later, when Barkev was able to visit, a bare week before her death, she complained to him, "Cathy thinks I'm dying but she's wrong."

The weeks that she was in the hospital were weeks when intravenous feeding kept her alive, retarding the ordinary rhythm by which life might slip away, until the cancer could destroy her more directly. She was very weak when she entered, thin and wasted and rejecting almost all food. In the natural course of things I suppose she would have died as almost all the creatures in nature die that do not die violently, in the slowly increasing weakness of those no longer able to take nourishment. Still, she was clear and insistent on the battle to retain life from one day to next. All that she had said about including death in life and being able purposefully to acknowledge the time of death was lost in the insistence on the continuing battle.

Margaret's involvement with the healer had begun in the same intellectual curiosity that made her insist through the years that one should at least look at the evidence for ESP and such phenomena. It seems to have evolved later into a replacement for the activism that goes with the invocation of heroic medical measures, a way of feeling from day to day that something at least is being done about the disease. It takes a special kind of courage, particularly rare in Americans, simply to await death. Thousands of dollars are expended in this country on treatments no more efficacious in holding back death than were the ministrations of the healer. Doctors seem as restless in their wish to tinker with the body during the process of dying as are patients and their relatives, desperate to feel that every possible effort has been made, filling time with the stresses associated with treatment. We have the courage of activity, but rarely the courage of passivity. Better to direct the impulse to do something into a ritual of "healing" that neither cut nor poisoned, brought neither nausea or pain. Unlike the servants of medical technology who tend to present a distant and impersonal image, the healer was a warm, broad-breasted and amusing woman, an intriguing personality combining different cultural traditions, ready to touch and to comfort.

Margaret even had the notion of doing a study of healers and of various forms of spiritual healing, just as Gregory, coming out of intensive care after his lung cancer was seen to be inoperable, spoke of writing his next book about the American medical establishment and about iatrogenic disease, disease caused by doctors. This is, after all, the traditional response in my family. When your life is lived in a double mode of participation and observation, then boredom, irritation, and even pain can be muted by the effort to understand the cultural patterns embedded in a given situation. On the other hand, it is not easy to sustain multiple levels of experience. Barkev and I worried that Margaret had gradually lost the ability to balance the insistence that she was not dying against the knowledge that in fact she was, making herself, by self-deception, vulnerable to deception and exploitation. In the last days before her death, I am told, Margaret realized that she was indeed dying and was angry, especially at the healer who had kept promising healing and found herself suddenly busy with other engagements at the moment of hopelessness.

I was back in Iran when I heard of Margaret's death. The disorder there had already reached the point where it was impossible to leave the country quickly enough to reach New York, and those around us were preoccupied by other deaths and the fear of increasing violence. We had a small memorial service in Tehran. The funeral itself was at St. Paul's Church, the chapel of Columbia University, in New York. She was cremated as she had wanted and indeed friends and family would have found unbearable the idea of that frail and wasted body laid in the winter earth, and then she was buried in the churchyard of the little Episcopal Church she had attended as a girl in Buckingham, Pennsylvania.

Later, in ordering her tombstone, I asked that it match the others around it as closely as possible, with only her name and the dates of birth and death over a phrase of her own that recurs over thirty years of her writing, "To cherish the life of the world." When Luther Cressman, her first husband, wanted to plant a rose bush by the grave, the church authorities declined because of the pruning that would be needed. She died within a framework of American tradition, spending her last weeks in a hospital and finally buried in a country graveyard, awarded posthumously the Presidential Medal of Freedom.

The manner of Gregory's dying in 1980 was very different, for he and his family had all had a two-year period of knowing that his death might come at any moment, a period when he had already been written off by conventional medicine. Lois, caring for him at Esalen, began to feel early in the spring an ebbing of vitality as he moved around less and less. Although his lung cancer remained in remission, he developed pneumonia and sharp pains in his side that we associated with the cancer. These pains were not identified as shingles until after a hospitalization for the pneumonia and until after he was very weak and blurred with drugs. From day to day he became more intolerant of hospital care and impatient for death, puzzled about the nature of the letting go that would release him.

After two weeks in the hospital, Lois made the decision to take him out and the San Francisco Zen community put us all up in their guesthouse. The Zen students shared with us in the tasks of care and sat in meditation near his bed, breathing in rhythm, around the clock. There were no tubes attached to Gregory's body, and when he ceased to accept food and brushed away the oxygen from his nostrils, we did not insist. I was there for most of the period of his illness, still not fully involved in new responsibilities after returning from Iran. The tranquillity of the Zen Center and the repetitive tasks of caring for Gregory were healing, as we cleaned and turned his great recalcitrant body.

Within the context of the careful detail of Zen attention, we each found our own small rituals of greeting and farewell. Watching alone with him in the dark and tranquil house during the last night before he died, when he still could press a hand and acknowledge loving presence but no longer spoke, I read aloud the passages he loved from the Book of Job, where God speaks of the mysterious wonders of the natural world, and held up a flower as one might hold up a cross to the gaze of a dying Christian, the only expression I could find, other than our own bodies, of the natural order he revered. Gradually, for another half day, his breathing slowed, and then it ceased. Lois closed his eyes and we bathed and dressed him and continued to take turns watching by his side for two more days, as others visited and the sense of his presence slowly receded.

After three days, our Zen friends went with us to the crematorium, into the actual room with the ovens, and we piled his body with wild flowers and parting gifts, a bagel from Nora that referred back to a breakfast-table conversation at Esalen, and a crab that John and Eric, his son and stepson, captured in the San Francisco Bay the night before, symbol of the dearest moments of childhood and of all the "fearful symmetries" of mind and nature. Our Zen friends chanted. Then they guided Lois to press the control on the great oven, and we went outside where we could stand in a meadow watching the line of smoke rising to the sky.

XV
Epilogue: And Part and Meet Again

The voices of my parents are still very much with me, for I hear their echo in so much that I see and encounter. They affirmed that what is most worth caring about is an interweaving of pattern going beyond any individual person, and their voices are blended now into the complex skein. It was difficult for me sometimes, as a child, to decode the differences and similarities and to pitch my own voice in harmony but yet distinct. The contained world of early childhood no longer exists, but my concerns remain similar to theirs and the analogies that bridge from the microcosm to the wider world continue, from Pere Village to New York, from the distorted communication of a schizophrenic family to the Pentagon. Finally, it is because of this play of analogy that I have felt that my own memories should be shared.

Margaret and Gregory are claimed by many people, not only by those who had an actual share in their lives, but by many others who never met them but have only read their writings or listened to a recorded voice or viewed an image on a screen. I find myself with a certain continuing responsibility for their writings and ideas, even as I acknowledge that my parents belonged to me no more than the photographs and records of my childhood, made in the specific so that someday that specificity could be projected out, an offering of enriching or clarifying models.

Increasingly as time passes, Margaret's and Gregory's names and faces are appropriated by groups and individuals to represent ideas they care about, whether with adulation or with opprobrium, but this process had already begun during their lives. In the sixties, the students told Margaret, "You belong to us; you really are one of us," but she said no, however responsive and sympathetic, her point of view was inevitably different. Buddhists in Boulder, Colorado, and followers of the Hindu holy man Sri Satya Sai Baba in Bangalore said to Gregory, "At heart you are really one of us," and laughed uneasily at the comments he made about their various supernaturalisms. In Manus, they called Margaret *pilapan*, female chief, and mother.

At Esalen, Gregory played the role of skeptic. A psychic came who painted pictures in the styles and names of the great Impressionist painters, painting that, like most of what is produced by psychics in the name of the mighty dead, were notably mediocre. To the horror of the community a small child defaced a painting by "Monet." Gregory defended her, teasing that this was a far stronger demonstration of the continued existence of the soul, since clearly the ghost of Monet had possessed the child in order to defend himself from a forgery. But after Gregory's death, the Esalen management published a whole catalogue in which photographs and previously unpublished writings by Gregory alternated with advertisements for their programs. Even as he was chiding and teasing, they felt he stood for an affirmation of what they cared about.

Margaret's name has been attached to a public school and to a patch of green near the American Museum of Natural History, a girl's dormitory at a halfway house for delinquents and an ecological hope ship, to name only a few. Gregory's name has been attached to a solar office building in Sacramento, and an institute for family therapy in France. Between the two of them there must have been at least twenty memorial services. Margaret was mourned at an extraordinary number of institutions: at the National Cathedral, at Columbia University, at the United Nations, at the American Museum of Natural History. Gregory was mourned in an extraordinary array of liturgical forms—Tibetan and Zen Buddhist, Hindu, Christian. Each of these symbolic acts is the staking of a claim to a share in meaning.

When Barkev and I were married, Margaret made it clear that our wedding was not a private occasion only, for such passages act as symbols in other people's lives. She added to the invitation list the names of many people I had never met or heard of, saying that this was her opportunity to express a continuing relationship to colleagues and childhood friends who had heard about me over the years. I remember going into the powder room to detach my veil which had come wildly askew and giving to someone that vague smile one gives to strangers seen reflected in the same mirror—and then realizing with a start that that remote look had been given to a guest at my own wedding. The first wedding I ever went to was the wedding of Moana Holt, whom as a baby my mother had held in her arms as they sheltered in a concrete water tank from a hurricane in Samoa, and Moana gave me the veil of the candy bride on her wedding cake, so a part of that wedding was mine. Then when I married, remembering that moment, I gathered all the small children together and had a picture taken with and for them, in case this was their first wedding. There was a couple among our friends who revived a dying relationship around our wedding and another who broke up, and one set of guests took a big bag of food to a Kennedy campaign gathering that night in the Bronx. The ceremony belonged to many people, as today many people lay claim to a share in my parents' passing.

We talk in this country often about property rights, but we talk more rarely about the shares people have in each other's lives, and about people's rights to participation and pleasure, especially at the moments of passage: the right to throw a handful of earth on a coffin, the right to stand up to catch a tossed bouquet and dream of one's own future wedding, to kiss a bride or groom and to hold a newborn. Couples today devise new rituals or set up housekeeping together in ways most meaningful to themselves without wondering whether meaning is something they owe to a larger community.

When we remember the dead, we often speak of how much they gave of themselves, usually thinking of the ways in which they have labored to build and to comfort. My parents gave of themselves in a different way as well, in the roles they played in the imaginations of so many people. I once asked Margaret why she didn't feel it was exploitative to accept so much from others, particularly from the adults who participated in my care, but also from the friends she would stay with as she traveled across the country who never came to visit us or those who gave dinners we did not reciprocate. She replied that she was letting people share in her life by feeling that they helped to make it possible, supporting the excitement and the achievement, and that the idea of reciprocating in kind was a bore. In this kind of self-giving she was endlessly generous. Only late in her life did she begin to put up the protective barriers that most people put up against phone calls and requests for interviews, and that was after it became fashionable for schoolteachers to tell children to "write to a famous person and ask them . . ." Gregory never put up effective filters, and at his various homes, especially in the last decade of his life, the hours were consumed by people, often strangers, dropping in to ask him questions about his work. Each, in his or her chosen fashion, was available to others, creating a community of expectation.

These expectations continue. Three years have passed since Gregory's death and almost five since Margaret's, but the legal estates are only barely through the complex process of settlement and Margaret's papers in the Library of Congress and Gregory's at the University of California at Santa Cruz are not yet fully catalogued. My father's final manuscript remains for me to complete as he wished, and requests from biographers, textbook writers, and filmmakers continue to come in. I could easily devote myself full time to responding to these requests and to the claims of those who want some important project strengthened by being associated with my parents' names.

What then are the obligations laid on me by the rights that others feel they have? What do I do when the notorious star of a TV series decides that she was born to portray Margaret Mead on television, or when a magazine of New Age spirituality decides to write on Gregory Bateson's religious beliefs on the basis of a telephone interview with me? How do I respond to the biographers who assume that their interest in one of my parents gives them a natural right to my time? Do the ethics change when the second such writer comes along, or the third, and does it make a difference if they have studied anthropology or knew Margaret or Gregory in their lifetime? The second movie proposal? The third? And is it my task to deal with attacks and criticism as they come along, some of them constructive and some petty and destructive, when friends and relatives call late at night pouring forth their grief and anger?

The poem my mother wrote for me in 1947, began

> *That I be not a restless ghost*
> *Who haunts your footsteps as they*
> *pass . . .*[1]

In fact, this poem is not about the kind of haunting involved in dealing with a complex legacy, but rather about the haunting that would have resulted from the attempt to determine a particular course to my life, but much of my life in this past five years has indeed been haunted by the need to deal responsibly with such questions.

In February 1983, the papers and news magazines were suddenly filled with reports on the work of an Australian anthropologist, Derek Freeman. Freeman claimed that my mother's work on Samoa, which emphasized the placidity of the culture and the relaxed attitudes toward sexuality that eased the maturation of the young girls, erred fundamentally in failing to recognize the violence and puritanism of the Samoans. He went on to use this attack on her first fieldwork, fifty years ago, as if it could itself constitute a refutation of all the careful work of many anthropologists over half a century in demonstrating the ways in which the behavioral differences between human groups depend on culturally patterned learning. This curious *ad hominem* attack on a body of theory was complicated by a systematic misstatement of that theory, suggesting that cultural anthropologists had somehow entirely dismissed biology as a factor in human behavior.

Preoccupied with another set of tasks, I hoped at first to stay out of the debate and leave it to those more expert on Samoa. Soon it became clear that because of the sensational way the Harvard University Press had chosen to promote the book, the issue went outside scholarship, and that my concern would be with defending not the details of my mother's first fieldwork but the ideas that were critical throughout her career and remain critical for me. It is precisely because the biological characteristics of human beings make learning possible that we are able, cumulatively, to develop rich and complex cultures and, finally, to make choices for the future.

It is logical to suppose that just as many people have wanted to build careers around an interest in Mead or Bateson, there will be those who see opportunity in an attack. The wish to personify the ideas you hate is only the mirror image of the hagiography that creates patron saints. It was after all Margaret's own decision to embed her name and image so widely in the thought of ordinary people, but this means that the debate was not limited to scientific issues but became rapidly politicized, exploited as an occasion to attack a range of liberal beliefs.

It is easy enough for me to visualize how resentment of working in my mother's shadow could have become an obsession for Derek Freeman. For forty years, every time he mentioned working in Samoa, he would have encountered comments about *Coming of Age in Samoa*. Since he worked at a different time and place and his initial viewpoint was gained from talking to chiefs, he inevitably developed a different emphasis, and over forty years of brooding his picture developed into the opposite of hers. The personality he projects is equally antithetical to hers—rigid and grim and competitive.

Freeman failed to identify the theoretical source of the flaws in *Coming of Age*, which is not the affirmation of the role of culture in patterning behavior but the expectation, at that stage in the development of anthropology, of a pervasive kind of homogeneity. Margaret was working then in the context of Ruth Benedict's developing ideas about cultural patterning, which led her to believe she would be able to characterize the style of the culture as a whole. In doing this, she allowed her sense of Samoa to be shaped by her initial intensive work with the adolescent girls in a remote and tranquil backwater. It is as if an entire picture of American culture were filtered through the vision of a group of teen-age girls in a small town before the invention of television.

This was no different, of course, from what male anthropologists had done routinely for years when they constructed cultural descriptions based primarily on interviews with important men—a partial and biased view at best. In fact, the ideas about adolescence in our own culture with which Margaret compared her data were based almost entirely on studies of males.[2] It was not until Margaret and Reo met Gregory in New Guinea, on the Sepik, that they began to develop the techniques for thinking about patterned contrasts, the counterpointing of placidity and violence or puritanism and permissiveness that can exist within a single culture, and the ways in which male and female points of view can be shaped to complement each other.

The discrepancies that Freeman found are comparable to those that female ethnographers have discovered in many places where their predecessors had believed for years that the women cowered in fear of the bullroarers or the terrifying masks brought out by the members of the men's clubs while, as often as not, the women were fully aware of the deception and amiably allowed the men the pleasure of their mystification. The young girls in Samoa may have known things that their fathers did not and that their mothers had forgotten, about what amorous explorations occurred under the palm trees—and at the same time been blind to many of the preoccupations of their elders, preoccupations with matters of status and political power from which they were excluded.

Margaret had sufficient intuition about the limitations of her work in Samoa so that she never again did extended fieldwork except in a working partnership with a male fieldworker who would see the society from those vantage points inaccessible to her. This was the possibility she foresaw in her marriage to Reo Fortune, not only a man whose work could complement her own but someone with an edge of intensity and abrasiveness that came as a relief after the loneliness of the field and the affability and relaxed good humor of the Samoan girls which finally she found cloying. In describing the Samoan ethos, she suggested that the very mental health she admired was bought at the cost of intensity and creativity. It has been important to me in thinking about the ways in which people have tried to distort the memory of my parents to a single hue, to remember that they valued debate and contrast.

The Freeman episode left me with the feeling about him, "There but for the grace of God go I." However, the only possible answer to his kind of criticism is a mixed answer, an attempt to move beyond the back and forth of adulation and attack, not a simple defense. At the same time, the debate discovered in me passions that I had not known I possessed. Freeman, fueled by accumulated venom, attacked and distorted ideas that I really do care about defending, ideas central to cultural anthropology that come to me from both parents. When we met on television, Freeman leaned over during a commercial break to sneer, "You really are a *pukka* cultural anthropologist, aren't you?" Indeed I am. Solid, clear through. Different as Margaret and Gregory were in their styles of thought and work, there was a common ground from which they moved, and both wanted to forge tools of understanding that would be liberating. My own course has been sufficiently different so that it has been surprising to me to find myself so clearly expressing allegiance to that common ground and to the continuing work, as if these might indeed provide the clarity that I looked for in different ways as I grew up, trying to savor commitment to ethnic and religious traditions.

In October 1978, when Vanni and I had come back to the United States to visit Margaret in the hospital, it was clearly her wish that I would be free to leave and to "get on with my work." At present, a reasoned and responsible stewardship of Margaret's and Gregory's intellectual legacy is an unavoidable task, rather in the sense that the study of the Arapesh became Margaret's task when the native bearers Reo had hired from farther inland abandoned them on a mountainside with six months' supplies and no way to move on to the richer culture they had hoped to study. There simply are tasks in relation to my parents' legacies that only I can do. The problem has been to find ways of carrying them out that will feel right to me and not to let them become the consuming preoccupations of a lifetime.

The decision to write this book was an attempt to define and delimit a portion of the task. Because I have not wanted to make myself available to biographers, I have felt an obligation to put my experience down in my own words. These will seem strange and alien enough when they are quoted later, worked into the interpretations of others, but at least I will be able once to phrase and tune them myself; I know too that they will seem strange and alien to others who cherish a different set of memories. In writing this book, I have deliberately not done those things that can be undertaken by other writers, scholars, or journalists, now or in the future. I have not conducted interviews, to add to the impositions on the time of others, though this book has inevitably been affected by conversations that were part of my continuing life while I worked on it. I have not worked my way through vast archives of correspondence, though I have had to be involved in effort and planning to be sure that the papers are responsibly handled and made available to the scholars who will want to use them—no doubt they will be able to correct many of my interpretations and memories, but perhaps my errors will be illuminating. I have not read or reread all the published works of Mead and Bateson, neither of which I know completely, for to do so would be to remake myself as an expert instead of a daughter. I have wanted to write a small book, a small part of the emerging library. I have tried to weave my own ambivalence into this book, letting love and grief, longing and anger, lie close to the surface, and making it clear that there is no perfection to enshrine and no orthodoxy to defend but much to use and much to value.

I played a curious and idiosyncratic role in the work of

both parents. Gregory invented a voice, the "Daughter" of the metalogues, in which I learned to speak, so that the actual dialogues and conversations we had in the last decade of his life were an important part of his ability to make his thought available to a wider community. He would happily have drawn me away from my other work and my other lives into sustained conversation with him about ideas, the only thing that really seemed to him worth doing, but the ways in which I can speak of his ideas now that the dialogue is over will be different.

For Margaret I was a colleague, one of many with whom she carried on her complex multilogue. It seems to me however that because so much of her thought was expressed in my upbringing, there is a sense in which my own happiness has had to carry the burden of proof for many of her ideas. The photographs published of Margaret over the years as a smiling mother or grandmother, meeting a responsive and echoing smile, have been dissertations, declarations that it is indeed possible to raise children so that they will be strong and joyous and creative and, in turn, work to build a better world. These are affirmations in which I join, even though no one is uniformly strong and joyous and there is no creativity without conflict. When people ask, "How does it feel to be the daughter of Margaret Mead?" they are asking a legitimate question, whose answer may be relevant to their own lives, but it will be nice to have set the answer down on paper.

I end this writing surrounded by echoes in patterns at least four generations deep. Even the poetry Gregory quoted repeatedly goes back several generations. One quotation from Samuel Butler, who also wrote on evolution, expresses my sense of the continuing conversation: "Yet meet we shall, and part, and meet again / Where dead men meet, on lips of living men."[3] William Bateson, pioneer in genetics, collected his own genealogy, and so did Margaret, watching for the prefigurations of their own intellect and the occasional instances of insanity and alcoholism they found spaced out over those generations. Gregory worried about whether the continuity with his family tradition that Margaret was concerned to preserve would really be a good thing for his children or a burden.

The choice between art and science, too, echoes for me from both traditions—my mother's sister Elizabeth becoming a painter, the suicide of my father's brother Martin, the piety those agnostic forebears of mine showed in relation to art. The genealogical explorations mirror the notebooks in which three generations of mothers have kept notes on the early development of their daughters in the reiterated conviction that we shall know even as we also are known, the choice of the personal science in which somehow the perception of the artist is also expressed. I find myself using phrases Margaret might have used and not knowing whether I am quoting.

However disparate the intellectual styles of Margaret and Gregory, it is important after so many chapters of seeking counterpoint to underline the themes on which they concurred, the themes which remain most important for me as well. For both of them the recognition of pattern was the overwhelming intellectual pleasure and the transmission and elaboration of evolving pattern the process most to be protected in both the biological and the social worlds.

In retrospect, we may find that in a few generations we value even more than their theoretical and methodological contributions the records they collected and preserved of human diversity. The ethnographic work done before World

War II by Margaret and Reo and Gregory and a few dozen other anthropologists will someday be our only way of recapturing the range of cultural variation lived by small preliterate human groups before they were intensively exposed to complex cultures and technologies. When everyone hums the same popular tunes, and dreams in the same seven-tone scale, and looks around at a world in which only the very sophisticated believe there are more than two alternative belief systems—and even they find their imagination limited to half a dozen—the records of the early fieldworkers will be brought out and studied for the rediscovery of human freedom and creativity. When Gregory was invited to comment on the pressure to teach "scientific creationism" in the California state school system, he shifted the ground of debate by telling of the Iatmul culture hero, Kevembuangga, who speared the crocodile whose churning tail maintained chaos, so that the land and water naturally separated, an inversion of the biblical notion that order is imposed from without. It was the diverse adaptation of the Galápagos finches that gave Darwin the clue for thinking about evolution, the clue that the order of species has not been divinely fixed for all time.

Margaret's thinking was more focused on the individual than Gregory's, in whose vision of the ecology of mind the envelope of skin around an organism is no more than a textural variation in the pattern of information transfer and cybernetic control. But both of them saw that the viability of the individual is embedded in ongoing communication, and therefore worried about protecting the systems in which such communications could take place. This means a concern for the redwood forests and the prairies and the ocean currents, as well as a concern for villagers in New Guinea, for New York City, for elementary schools, and for small groups of thinkers exchanging and developing ideas in a conference room—"passing on programs," as Gregory said at Burg Wartenstein.

Margaret felt far more able to act than Gregory, partly out of temperamental difference and partly, perhaps, because of the contrast in their wartime experience. The two violations of systemic pattern that frightened Gregory the most were the distortion of communication and the elimination of diversity, since internal diversity and multiplicity of complementary pathways for information are what allows a system to survive in the face of change. His war work involved the introduction of misinformation and the violation of patterns of communication in the hope of thereby damaging the enemy.

Margaret's war work was the other face of applied anthropology: She was involved in enriching the communicative networks available in our own society—enriching the efforts at allied cooperation by supplying each side with a cultural perspective on the differences between them, or supplying planners with a sense of cultural preference in nutrition as well as biological needs. Thus she felt able to build and change by supplying a sense of alternatives, creating a diversity of institutional form, and increasing self-awareness. Gregory felt, in considering human problems, that there were errors, wrong premises that might be abandoned or rooted out in some unstated way. Margaret felt, I believe, that by supplying new metaphors and additional layers of insight, errors would be balanced.

She once said to me, talking about Larry Frank, that the trouble with people who reject religion is often that they go on believing in the devil after they give up God, that too many intellectuals, my father among them, are oppressed by a sense of evil without a confidence in good. At the end of his life Gregory's sense of the integrity and pervasiveness of pattern in the biosphere had developed a religious vision, but he was still groping for ways of acting within that vision that would not violate its sacred unity. Knowledge remained for my parents the basis of both action and ethical commitment. Gregory set out once to define love. "You could say," he said, "that at least a part of what we mean by the word could be covered by saying that 'I love X' could be spelled out as: 'I regard myself as a system, whatever that might mean, and I accept with positive valuation the fact that I am one, preferring to be one rather than fall to pieces and die; and I regard the person whom I love as systemic; and I regard my system and his or her system as together constituting a larger system with some degree of conformability within itself.' "[4]

This is a definition of relationship as knowledge, achieved and sustained through information exchange—through conversation and communion, whether at the level of genetics or of government. We are still learning how to recognize and protect systemic patterns of communication, to act on the wisdom of recognition, behave as if we are parts of a single whole. In the face of the last surviving member of an Indian tribe, in the face of a Soviet premier or a child, we see the human lineaments, the marks of learning and the possibility of further learning. The wisdom of recognition and participation is the same as we look at the contours of a field or the changing face of a living lake.

For me, the sense of recognition is now layered. Not only do I try to recognize the congruence of other living systems, I also see the faces and hear the voices of my parents. As time goes on, I will speak and write of them less often, moving to other subjects and modes of expression, but I will still be exploring many of the same ideas. Even as we discard some of their formulations as imperfect or incomplete, we will draw on their ideas while we struggle as a species to avoid irreversible destruction. I have wanted to retain and perhaps to share the specificity of touch and vision, to listen to their voices as we speak to each other about them, to go on meeting them. I wish too, because so much that each of them wanted to teach went beyond what they could express fully as scientists, that we could go on meeting them, as we continue to think about the issues that concerned them, as artists and lovers as well, for their knowledge was based in caring and, in different ways, their books are full of poetry.

References and Sources

This is a book based primarily on experience. After some consideration, I have decided to keep the formal paraphernalia of scholarship to a minimum. References for actual quotations from printed sources (original editions unless otherwise noted) are provided for each chapter, but I have not attempted to find published references for material that entered my knowledge through conversation, although many such exist, nor have I given large numbers of footnotes to "personal communication," which seem to me useful only when there has been systematic interviewing and dated interview notes exist. The reader should assume that quotations for which references are not given are reconstructed from memory and may not be absolutely exact.

There may be readers who wish to pursue some subjects further, so I have mentioned a few works important for background before the references to certain chapters. These are gathered in a selective bibliography along with works referred to repeatedly, especially my mother's autobiography

(Mead 1972), the biography of my father by David Lipset (1980), and the published bibliography of each (Gordan 1976 and Levi and Rappaport 1982)—all being important resources. These works are cited in the specific references only by author and date. Works referred to in passing, like Shakespeare's plays or children's books read to me as a child, are not fully referenced.

After some thought, I decided that this memoir should not have an index, lest it acquire thereby a spurious air of systematicity. The logical mix of topics resists alphabetization. The selection of individuals referred to does not reflect their importance in the lives of my parents but only their relevance to the discussion of some aspect of our relationship.

Selected Bibliography

BATESON, GREGORY

1936   *Naven: A Survey of the Problems Suggested by a Composite Picture of the Culture of a New Guinea Tribe Drawn from Three Points of View.* Cambridge, England: Cambridge University Press. Revised edition, Stanford University Press, 1958.

1972   *Steps to an Ecology of Mind.* San Francisco: Chandler Press.

1979   *Mind and Nature: A Necessary Unity.* New York: E. P. Dutton.

———WITH MARGARET MEAD

1942   *Balinese Character: A Photographic Analysis.* New York: New York Academy of Sciences.

———WITH JURGEN RUESCH

1951   *Communication: The Social Matrix of Psychiatry.* New York: Norton.

BATESON, MARY CATHERINE

1972 *Our Own Metaphor: A Personal Account of a Conference on Conscious Purpose and Human Adaptation.* New York: Knopf.

BENEDICT, RUTH

1934 *Patterns of Culture.* Boston: Houghton Mifflin.

1946 *The Chrysanthemum and the Sword: Patterns of Japanese Culture.* Boston: Houghton Mifflin.

ERIKSON, E. H.

1950 *Childhood and Society.* New York: Norton.

FORTUNE, REO

1935 *Manus Religion.* Philadelphia: American Philosophical Society.

FREEMAN, DEREK

1983 *Margaret Mead and Samoa: The Making and Unmaking of an Anthropological Myth.* Cambridge, Mass.: Harvard University Press.

GORDAN, JOAN, ED.

1976 *Margaret Mead: The Complete Bibliography 1925–1975.* The Hague: Mouton.

LEVI, ROBERT I., AND ROY RAPPAPORT

1982 *"Obituary: Gregory Bateson, 1904–1980," American Anthropologist,* Vol. 84, No. 2:379–394.

LIPSET, DAVID

1980 *Gregory Bateson: The Legacy of a Scientist.* Englewood Cliffs, N.J.: Prentice-Hall.

MEAD, MARGARET

1928 *Coming of Age in Samoa: A Psychological Study of Primitive Youth for Western Civilization.* New York: Morrow.

1930 *Growing up in New Guinea: A Comparative Study of Primitive Education.* New York: Morrow.

1935 *Sex and Temperament in Three Primitive Societies.* New York: Morrow.

1956 *New Lives for Old: Cultural Transformation—Manus, 1928–53.* New York: Morrow.

1959 *An Anthropologist at Work: Writings of Ruth Benedict.* Boston: Houghton Mifflin.

1964 *Continuities in Cultural Evolution.* New Haven: Yale University Press.

1970 *Culture and Commitment: A Study of the Generation Gap.* Garden City, N.Y.: Natural History Press/Doubleday.

1972 *Blackberry Winter: My Earlier Years.* New York: Morrow.

1974 *Ruth Benedict.* New York: Columbia University Press.

1977 *Letters from the Field, 1925–1975.* New York: Harper & Row.

——WITH JAMES BALDWIN
1971 *A Rap on Race.* Philadelphia: Lippincott.

——WITH PAUL BYERS
1968 *The Small Conference: An Innovation in Communication.* The Hague: Mouton.

——WITH RHODA METRAUX
1980 *Aspects of the Present.* New York: Morrow.

MODELL, JUDITH S.

1983   *Ruth Benedict: Patterns of a Life.* Philadelphia: University of Pennsylvania Press.

OLMSTED, D. L., ED.

1980   *"In Memoriam: Margaret Mead (1901–1978),"* *American Anthropologist,* Vol. 82, No. 2.

SAPIR, EDWARD

1921   *Language: An Introduction to the Study of Speech.* New York: Harcourt Brace.

SCHWARTZ, THEODORE

1962   *The Paliau Movement in the Admiralty Islands, 1946–1954.* Anthropological Papers of the American Museum of Natural History, 49, Pt. 2. New York.

Notes

I. THE AQUARIUM AND THE GLOBE

1. Margaret Lowenfeld, *The World Technique* (London: The Institute of Child Psychology Press, 1970).

II. BABY PICTURES

For extended descriptions of the family backgrounds and youths of Margaret and Gregory, see Mead 1972 and Lipset 1980.

1. Mead 1972: 263.

2. Mead, "Of Mothers and Mothering: Remembrances," *The Lactation Review*, Vol. IV, No. 1: 6.

3. Benjamin Spock, *Baby and Child Care*. Third edition (New York: Hawthorn Books, 1968), p. 61.

4. Mead 1972: 268.

5. Mead 1980: 140–146.

III. A HOUSEHOLD COMMON AND UNCOMMON

1. Mead 1972: 80.

IV. "DADDY, TEACH ME SOMETHING"

For two decades after the publication of his book on psychiatry (G. Bateson and Ruesch 1951), Gregory published primarily in article form. Those interested in pursuing his work on play or schizophrenia or other topics mentioned here can find the principal pieces anthologized in G. Bateson 1979.

1. This story is told in "Epidemiology of Schizophrenia" (1955, reprinted in G. Bateson 1972: 198–199) about gladiolas, but I remember it of chrysanthemums.

2. Reprinted in C. G. Jung, *Memories, Dreams, Reflections: 1875–1961*, Aniela Jaffé, ed. (New York: Pantheon, 1963).

V. COMING OF AGE IN NEW YORK

1. Margaret's shift from believing that children could simply ignore race is discussed in her joint book with James Baldwin (1971).

VI. ONE WHITE GLOVE AND THE SOUND OF ONE HAND CLAPPING

1. Draft MS. for Mead 1972, Library of Congress Collection, 2/24/71, p. 11.

2. Mead 1928: 33.

3. William Blake, "Jerusalem," sec. 55.

4. e. e. cummings, "Spring is like a perhaps hand."

5. Eugen Herrigel, *Zen in the Art of Archery* (London: Routledge, 1953).

VII. AWAY FROM THIS FAMILIAR LAND

1. For instance, Lawrence and Mary Frank, *How to Help Your Child in School* (New York: Viking Press, 1950), and

How to Be a Woman: From Girlhood Through Maturity (New York: Bobbs-Merrill, 1954).

2. M. C. Bateson, *Structural Continuity in Poetry: A Linguistic Study of Five Pre-Islamic Arabic Odes* (The Hague: Mouton, 1970).

3. Mead 1982: 272.

VIII. Sharing a Life

Margaret herself published two biographies of Ruth Benedict (1959 and 1974). A more extensive biography appeared while this book was in draft (Modell 1983).

1. Mead and Metraux 1980: 269–275.

2. Mead 1959: 56.

3. Ibid.: 58.

4. Ibid.: 89.

5. Mead 1972: 132.

IX. Sex and Temperament

The research on the Sepik to which this chapter refers is described in Mead 1935 and G. Bateson 1936 and other more technical works. It is also discussed in Mead 1972 and 1977, but I found that I had to go to the draft manuscript of Mead 1972 (in the Library of Congress) for clarification of many points.

1. Mead 1972: 220.

2. Deborah Gewertz, *Sepik River Societies: A Historical Ethnography of the Chambri and Their Neighbors* (New Haven: Yale University Press, 1983).

3. Mead 1972, pp. 216–222 and sections of draft MS., dated January 1972.

4. Draft MS. for Mead 1972, 1/20/72, p. 37. Note that in the version of this anecdote given in the draft Margaret used

letters of the alphabet to refer to the types but I have re-instated the references to points of the compass which Reo Fortune would have used at the time.

5. Gregory's version of this chart conflicts with Margaret's. When we reviewed it he labeled Westerners as both caring and careful and Easterners as neither. It was Gregory who sent me back to the draft by discovering that in Mead 1972 East and West were reversed.

6. Mead 1972: 208.

7. Draft MS 1/20/72, p. 30.

8. Rudyard Kipling, "The Sons of Martha."

X. PARABLES

The postwar story of the Manus people was told by my mother in *New Lives for Old* (1956) and by Theodore Schwartz (1962). The prewar Manus are described in Mead 1930 and Fortune 1935.

XI. PARTICIPANT OBSERVERS

The family correspondence quoted in this chapter is cited more extensively by Lipset (1980), whose interpretation and selection have influenced my own. Gregory's own explanation of his use of the Theory of Logical Types of Russell and Whitehead and of cybernetics are most accessible in G. Bateson 1979.

1. Martin Bateson to his mother, October 31, 1917, quoted in Lipset 1980: 66.

2. William Bateson to Gregory, April 23, 1922, in Lipset 1980: 96.

3. Mead 1972: 107.

4. Ibid., 111.

5. Gregory to his parents, July 21, 1925, in Lipset 1980: 115.

6. Erik Erikson, "On the Nature of Clinical Evidence," *Insight and Responsibility* (New York: Norton, 1964).

7. Lipset 1980: 63.

8. This account actually conflates experiences both in the water and beside the pool with three different dolphins—Peter, Cissy, and Roger—in St. Thomas and Hawaii. I am no longer certain of the sequence.

9. "Mother-Infant Exchanges: The Epigenesis of Conversational Interaction," *Developmental Psycholinguistics and Communication Disorders*, 1975, ed. Doris Aaronson and Robert Rieber, Annals of the New York Academy of Sciences 263: 101–113.

10. "Insight in a Bicultural Context," *Philippine Studies*, 16, 4: 605–621.

XII. Our Own Metaphor

1. Arthur Koestler, *The Call Girls: A Tragicomedy with Prologue and Epilogue* (London: Hutchinson, 1972), and Arthur Koestler and J. R. Smythies, eds., *Beyond Reductionism: New Perspectives in the Life Sciences*, proceedings of the Alpbach Symposium (London: Hutchinson, 1969).

2. Lipset 1980: 262.

3. M. C. Bateson, 1972: 28.

4. Ibid.: 16.

5. G. Bateson 1979: 1.

6. G. Bateson, "The Double Bind—Misunderstood?" *Psychiatric News* 13 (1978): 40, quoted in Lipset 1980: 296.

XIII. Chère Collègue

The three works mentioned in this chapter as dealing with the generation and transmission of knowledge are Mead 1964 and 1970 and Mead and Byers 1968.

1. Edna St. Vincent Millay, "First Fig."
2. Ibid., "Recuerdo."
3. Rudyard Kipling, "When Earth's Last Picture Is Painted."
4. Mead 1928: 16.
5. T. A. Sebeok, A. S. Hayes, and M. C. Bateson, eds., *Approaches to Semiotics: Anthropology, Education, Linguistics, Psychiatry, Psychology*, Proceedings of the Indiana Conference on Paralanguage and Kinesics (The Hague: Mouton, 1964).

XIV. Steps to Death

The discussion of overlapping life cycles in this chapter owes much to Erikson 1950 and Mead 1964. The book on which I assisted Gregory is G. Bateson 1979.

1. Mead 1959: 438.

XV. Epilogue: And Part and Meet Again

1. Mead 1972: 272.
2. Carol Gilligan, *In a Different Voice: Psychological Theory and Women's Development* (Cambridge: Harvard University Press, 1982).
3. Samuel Butler, "The Life After Death."
4. M. C. Bateson 1972: 280.